Mastodon

by Chris Minnick and
Michael McCallister

Mastodon For Dummies®

Published by: **John Wiley & Sons, Inc.**, 111 River Street, Hoboken, NJ 07030-5774, www.wiley.com

Copyright © 2023 by John Wiley & Sons, Inc., Hoboken, New Jersey

Published simultaneously in Canada

No part of this publication may be reproduced, stored in a retrieval system or transmitted in any form or by any means, electronic, mechanical, photocopying, recording, scanning or otherwise, except as permitted under Sections 107 or 108 of the 1976 United States Copyright Act, without the prior written permission of the Publisher. Requests to the Publisher for permission should be addressed to the Permissions Department, John Wiley & Sons, Inc., 111 River Street, Hoboken, NJ 07030, (201) 748-6011, fax (201) 748-6008, or online at http://www.wiley.com/go/permissions.

Trademarks: Wiley, For Dummies, the Dummies Man logo, Dummies.com, Making Everything Easier, and related trade dress are trademarks or registered trademarks of John Wiley & Sons, Inc. and may not be used without written permission. Mastodon is a trademark of Mastodon GmbH. All other trademarks are the property of their respective owners. John Wiley & Sons, Inc. is not associated with any product or vendor mentioned in this book.

For general information on our other products and services, please contact our Customer Care Department within the U.S. at 877-762-2974, outside the U.S. at 317-572-3993, or fax 317-572-4002. For technical support, please visit https://hub.wiley.com/community/support/dummies.

Wiley publishes in a variety of print and electronic formats and by print-on-demand. Some material included with standard print versions of this book may not be included in e-books or in print-on-demand. If this book refers to media such as a CD or DVD that is not included in the version you purchased, you may download this material at http://booksupport.wiley.com. For more information about Wiley products, visit www.wiley.com.

Library of Congress Control Number: 2023930252

ISBN 978-1-394-19336-3 (pbk); ISBN 978-1-394-19338-7(ebk); ISBN 978-1-394-19337-0 (ebk)

SKY10041398_012023

Table of Contents

CHAPTER 7: Running Your Own Mastodon Instance 99

CHAPTER 8: Ten Tools that Integrate with Mastodon 125

Introduction

Welcome! Whether you're curious about Mastodon as an alternative to billionaire-owned corporate social media platforms or because you've noticed that many of your favorite people are joining Mastodon, this book is your friendly and easy-to-use step-by-step guide.

Maybe you're trying to decide whether to join Mastodon. Or maybe you've already joined Mastodon and are looking to get the most out of it. Or maybe you stumbled over some of the technical details of signing up on your first attempt but you're ready to give it another go. Whatever your reasons for wanting to learn more about Mastodon, we believe that you now hold in your hands (or on your screen) the easiest, friendliest, and most complete guide available!

For those of us (like Chris) who came to Mastodon because we had a sudden epiphany that we no longer liked Twitter, after spending way too much time on it, Mastodon feels like a breath of fresh air. But it's a different kind of fresh air — like a sea breeze at a beach full of interesting people you want to talk to, rather than angry knuckleheads, salespeople, and snake-oil peddlers.

For those of us (like Michael) who came to Mastodon in its early days because of the cool technology and its potential to change the way the web works for the better, the sudden popularity and success of Mastodon is proof that social media doesn't need to be corporate owned and that online communities can form and become sustainable without profit motive.

Thank you for choosing us to be your guides as you take your first steps into the world of Mastodon.

About This Book

This book is designed for people who are either new to Mastodon or have some experience but want to learn how to make better use of it. You'll discover, in plain English, the most important and useful things to know:

>> Learning the history of Mastodon

>> Understanding how Mastodon is different from and similar to Twitter

- ⟫ Knowing what it means that Mastodon is distributed
- ⟫ Signing up for a Mastodon account
- ⟫ Learning how to log into the Mastodon mobile app and the website
- ⟫ Customizing your Mastodon profile
- ⟫ Configuring important preferences in Mastodon
- ⟫ Getting a verified link on Mastodon
- ⟫ Securing your Mastodon account
- ⟫ Using Mastodon's three timelines
- ⟫ Searching for and finding other users
- ⟫ Following other users
- ⟫ Using hashtags on Mastodon
- ⟫ Viewing trending topics
- ⟫ Blocking users and servers
- ⟫ Getting followers
- ⟫ Interacting with people and posts
- ⟫ Using direct messages
- ⟫ Understanding how moderation works
- ⟫ Posting images and video
- ⟫ Creating polls
- ⟫ Doing business on Mastodon
- ⟫ Making positive contributions as a business on Mastodon
- ⟫ Setting up and running your own Mastodon website

As you read this book, keep the following in mind:

- ⟫ The book can be read from beginning to end, but feel free to skip around if you like. If a topic interests you, start there. You can always return to previous chapters, if necessary.

- ⟫ At some point, you'll get stuck, and something won't work as you expect. Do not fear! Many resources are available to help you, including Mastodon's documentation (at https://docs. joinmastodon.org/), other people on Mastodon, and us! You can send Chris Minnick a public message on Mastodon at

@chrisminnick@hachyderm.io and you can send Michael McCallister a public message at @workingwriter@fosstodon.org. Make sure to use the hashtag #mastodonFD!

In the book, you may note that some web addresses break across two lines of text. If you're reading this book in print and want to visit one of these web pages, simply key in the web address exactly as it's noted in the text, pretending as though the line break doesn't exist. If you're reading this as an e-book, you've got it easy – just click the web address to be taken directly to the web page.

Foolish Assumptions

We make only a few assumptions about you, the reader.

We assume that you have a computer running a modern web browser or a mobile device running Android or iOS. The instructions and screenshots in the book were tested and optimized using the Chrome browser, which is available for free from Google. Even so, all the instructions and websites mentioned will work in the latest version of Firefox, Safari, Microsoft Edge, and any other modern web browser.

We assume you have access to an internet connection. Just as with any social network or mobile app, you won't be able to do much with Mastodon if you're not connected to the internet.

Icons Used in This Book

Here are the icons used in the book to flag text that should be given extra attention or that can be skipped.

TIP

This icon flags useful information or explains a shortcut to help you understand a feature or concept.

REMEMBER

Try not to forget the material marked with this icon. It signals an important concept or process that you should keep in mind.

This icon explains technical details about the concept being explained. The details might be informative or interesting but are not essential to your understanding of the concept at this stage.

TECHNICAL STUFF

Watch out! This icon flags common mistakes and problems that can be avoided if you heed the warning.

WARNING

Beyond the Book

Extra content that you won't find in this book is available at www.dummies.com. Go online and type **Mastodon For Dummies Cheat Sheet** in the Search box to see the following:

>> **Cheat Sheet:** The online cheat sheet provides additional tips for using Mastodon and building your following.

>> **Updates:** Mastodon is growing and improving rapidly. As such, some of the instructions or screenshots in this book may not be exactly correct when you read this book. You'll find updates to the book, as well as a place to report errors that you may find.

Where to Go from Here

Are you ready to get started? Turn the page! Remember: You don't have to read this book in order from start to finish. If you're most interested in learning how to sign up for Mastodon, go straight to Chapter 2. If you want to know whether Mastodon would be a good fit for your business, check out Chapter 6. If you want to get technical and find out how to get started with setting up your own Mastodon server, go straight to Chapter 7.

No matter where you decide you're going to start reading this book, we're glad you are. And we're looking forward to seeing you on Mastodon!

Chapter **1**
Exploring Mastodon

You've probably heard that Mastodon is the open-source Twitter or a Twitter alternative. There's certainly some truth to those definitions, but they don't really tell you all that much about Mastodon itself.

It's true that Mastodon is a microblogging service like Twitter. You can post short messages of up to 500 characters that friends and strangers can read and respond to. You can share links to news, information, photos, and video from across the internet.

But Mastodon is far from being a Twitter clone. It operates on a different economic model and isn't owned by anyone, especially a celebrity billionaire.

In this chapter, you find out how Mastodon works in the broader scheme of things. We provide a more detailed comparison with Twitter, definitions of terms you'll encounter while exploring Mastodon, a little bit of history, and more. If you're anxious to get started using Mastodon, feel free to skip to the next chapter. We understand. Come back to this chapter when you're ready.

A Brief Definition of Mastodon

At the `https://mastodon.help` site, they answer the question "What is Mastodon?" this way:

A good way to describe the network is by saying that Mastodon is a 'galaxy of interconnected social networks based on a common platform.'

Mastodon isn't just one site, like Twitter.com and Facebook.com. The first choice you'll make when you decide to create an account is what star (or *instance*) in this galaxy of sites to call home. (Chapter 2 walks you through the steps.)

Each of these sites runs the Mastodon software (the *common platform*). Once you have a home on a site running the Mastodon software, you can connect with people on other sites running Mastodon.

However, there are exceptions to your ability to connect. Your site administrator can choose to block connections to some or all Mastodon instances for whatever reason. Most often, blocking happens because a particular site's users frequently harass other people, produce spam, or engage in some other bad behavior. We talk more about your ability to block individuals and servers in Chapter 3.

You may wonder where all these instances came from. Perhaps more to the point, how does Mastodon pay its bills? There are two parts to that answer.

Mastodon's code is open source, and the person who wrote the code doesn't charge anyone to use it. But just because the Mastodon software doesn't cost anything doesn't mean there aren't costs to running Mastodon. Internet service providers don't connect sites to the internet for free. The web servers that host the content on Mastodon cost money to operate and maintain. Parts break and need to be replaced. Larger instances serving thousands of users need more than a volunteer admin (or even a squad of volunteers) to keep everything going.

Twitter and other big tech companies solve this problem by selling targeted advertising to companies that want to reach specific types of people. Everything you do on their platforms is tracked: The posts you like, the posts you share, the topics you post on, and more. Many of them track all your travels on the internet through browser cookies. Your privacy doesn't matter to them.

Mastodon doesn't work like that. The vast majority of instances don't accept ads, and many block access from instances that do sell ads.

In your travels around Mastodon, you'll see many reminders to "tip your bartenders." The folks who manage Mastodon instances usually have a way for users to contribute to the instance's expenses, either directly, through a foundation, or through a platform such as Patreon or Kickstarter. The best administrators are transparent about expenses, posting the monthly expenses of the instance, and describe in some detail where your contributions go. Be generous if you can.

A Brief History of Mastodon

Mastodon was announced on HackerNews.com in October 2016 by its German creator, Eugen Rochko (gargron@mastodon.social). See Figure 1-1. It quickly became a sensation, in part because Twitter was thought to be in financial trouble, unable to sell enough advertising.

In its first two years, the Mastodon network grew to 1,627,557 registered users on 3,460 servers.

Many of those early users were people obsessed with technology and privacy, and the perceived difficulty of getting started led to some stagnation. But while other social media services, such as Google Plus, were disappearing, Mastodon continued plodding onward. Mastodon's slow-but-steady growth profile changed massively in April 2022, when Elon Musk started making noises about wanting to buy Twitter. When he succeeded in September, a race to the exits began, and millions came to explore Mastodon.

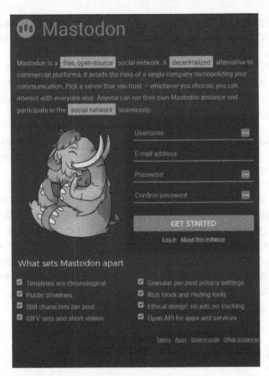

FIGURE 1-1: A Wayback Machine image of the mastodon. social about page on May 6, 2017.

In November 2022, Rochko announced that Mastodon had reached 1,028,362 monthly active users. By January 2023, Mastodon had 9 million registered users and 12,000 servers. The volume of posts on Mastodon had reached 500 million per month, according to the Fediverse Observer (https://mastodon.fediverse.observer/stats), a site that tracks various usage statistics for Mastodon and similar federated social sites. (See "Welcome to the Fediverse!" later in the chapter to learn more about the Fediverse.)

Comparing Mastodon to Twitter

In this section, we describe some of the main differences between Mastodon and Twitter.

On Mastodon, conversation is the focus

Patrick Hogan described Mastodon in 2017 as "a social media service where users actually feel comfortable being themselves, as opposed to a performative, more sarcastic version of who they actually are."

Many people arrive on Twitter looking to raise their public profile, gain millions of followers, and go viral. Certainly some people have become media pundits and celebrities by way of their Twitter posts.

On Mastodon, however, being sociable, informative, and authentic is more important. A Mastodon instance is better thought of as a neighborhood rather than a stage. You don't have to be smart-alecky to be respected.

In addition, before getting an account on a Mastodon instance, you sign on to a code of conduct. Although the code of conduct can go into detail regarding what is and isn't allowed, the code is often simply a lengthy version of "while you're on here, be nice." (You learn more about codes of conduct when you set up your account in Chapter 2.)

Mastodon is a home, not a site

If you want to tweet, you go to Twitter.com, get an account, agree to the terms of service (whether you read all that legalese or not), and start tweeting. Twitter may not ask much of you as a user, but you pay a price.

You have no stake in Twitter as a company, except as an unpaid content contributor. Twitter polls run by the CEO aren't binding and wouldn't be even if a majority of participants were stockholders.

Every Mastodon server instance is an island to itself but interconnected. Instance administrators and moderators are not gods and can't make any decisions beyond that instance. That's true even for founder Eugen Rochko.

The vast majority of Mastodon sites seek to be safe and friendly spaces for all kinds of users. You should feel like you're at home here.

Mastodon is built to serve its community

Mastodon is not driven by its financial bottom line nor the whims of venture capitalists, Wall Street traders, and hedge fund managers. It's driven to serve its users and the community.

If the founder or his heirs were to sell the mastodon.social site, where Mastodon was born, each user could easily move to another Mastodon instance, taking their posts and followers with them.

If you've found a home with a group of admins and moderators that serve you and the community at large, consider helping to keep the servers (and the community) running with a financial contribution that fits your budget.

Algorithms aren't allowed on Mastodon

Twitter started out as a place where you read the tweets of the people you followed in reverse-chronological order (as you do in blogs). But as the service grew, Twitter developed algorithms to figure out what you liked and shared on the service. The purpose of the algorithm was to keep you scrolling through your feed. You got what the algorithm thought you wanted, but you also saw more ads as the session continued. More eyeballs on ads, more money for Twitter.

As politicians and politically oriented people found a way to get followers, raise money, and gain influence on the service, ordinary users found themselves in a filter bubble, where they saw only things with which they agreed. Debate, where it existed at all, became toxic. Misinformation flourished. Some people started calling Twitter a hellscape because of all the angry tweets that the algorithm promoted into people's feeds.

No Twitter user ever got to opt-out of the algorithm's role in managing their feeds. There was never a vote among users to implement the algorithm when it was first developed. Programmers from outside the company couldn't review the algorithm to determine its effect on people's emotions.

In contrast, Mastodon displays every post from every user and hashtag you follow. You may still spend hours scrolling through interesting posts by fascinating people, but at least you know that Mastodon won't try to manipulate you to stay longer on the platform.

Users aren't tracked on Mastodon

The same forces (sometimes called *surveillance capitalism*) that created algorithms also created the cookie, that tiny slice of code that lives in your browser and tracks your movements across the platform and elsewhere. The cookie's owner aims to understand the things you're interested in or curious about — the better to sell your social graph to advertisers.

The Mastodon focus on privacy actively discourages cookies. Nearly every instance offers strict rules on such behavior, and should an instance permit it, that instance would likely be blocked by most other Mastodon instances.

Don't bet on this changing anytime soon.

Ads aren't acceptable (yet) on Mastodon

The initial burst of enthusiasm for Mastodon came when Twitter appeared to be placing its financial stability on selling more advertising on the platform. The Great Mastodon Migration of 2022, which came next, was made up of users and advertisers who didn't want to be associated with Twitter's new owner.

On Mastodon, most codes of conduct discourage excessive advertising, and what's excessive has mostly counted in single digits. The exceptions are the online equivalent of yard sales, art shows, community theater, and musical performances.

The Mastodon focus on community-building frowns on personal branding at least as much as on corporate branding. However, businesses aren't banned, and if you're in business, you can post about it all you want. (We offer tips for businesses to succeed on Mastodon in Chapter 6.) What won't fly are targeted ads based on trackers and algorithms, as noted previously. Mastodon isn't perfect, but its ideals are high.

Welcome to the Fediverse!

The *fediverse* is a collection of websites and social networks that in some ways looks back fondly on the early days of the World Wide Web. When Sir Tim Berners-Lee created the web, he envisioned it as a collaboration tool, with sites offering read-write access to any visitor. Aside from wiki sites (such as Wikipedia) and blogs that accept reader comments, today's web is largely a space for one-way communication.

Mastodon encourages its users to collaborate by putting the focus on conversation and discussion. It's easy to participate in a conversation by selecting the post's reply icon. The Fediverse expands that idea beyond just Mastodon into other services.

TIP

The fediverse, which is a combination of *federated* and *universe*, got its name for the federated connections between independent websites on the social web.

The foundation of the fediverse, and Mastodon as well, is the ActivityPub protocol, recognized as a web standard by the World Wide Web Consortium (W3C). Developers use the standard to power a variety of federated alternatives for organizing events, sharing music, and just hanging out. (You can learn more about ActivityPub at https://activitypub.rocks.)

TECHNICAL
STUFF

The boring, technical way to describe the ActivityPub standard is that it "provides a client to server API for creating, updating and deleting content, as well as a federated server to server API for delivering notifications and subscribing to content." Mastodon first used a protocol called OStatus, an open standard that enables microblogging, but switched to ActivityPub in 2017.

For the last decade or so, the IndieWeb movement (https://indieweb.com) has been building some of the social web standards that power Mastodon, especially the ActivityPub protocol, but for individual websites. IndieWeb focuses on your ability to control your content and connect with the people you want to connect with, with no one using your stuff to get others to sell you things you may not want.

Do not confuse *fediverse* with *metaverse!* The metaverse started as a creation in the mind of speculative fiction author Neal Stephenson in his fantastic novel *Snow Crash,* where it served as the escape hatch from the dystopia in which most people lived, where the only jobs were in high tech and high-speed pizza delivery. The dystopia was the society. Today, the metaverse is a virtual reality that Facebook (excuse us, Meta) billionaire Mark Zuckerberg wants to use to make even more money.

The fediverse is a much more pleasant place, where the users rule. And if profits are even a consideration, they come after the people who participate.

Understanding Federation

When you're starting out with Mastodon, wrapping your head around the idea of a *federation* can be maddening.

Not to overwhelm you with community metaphors, but think about the following facets of neighborhoods:

>> Life might be pretty much the same from neighborhood to neighborhood, but some places have different social rules. Violating those rules can get you in trouble. In some neighborhoods, you must live there for many years before you can even call yourself a resident!

>> Although your neighborhood might have regular elections for city council members and other leaders, you might also have a person active in the community who is so respected that they're called the mayor of the neighborhood.

>> If you have an active neighborhood association, they probably have events to raise money for various programs to improve life in the neighborhood. While the city government pays for a lot of things (and usually has taxing power), neighborhood associations often pay for things such as flowers in common areas.

>> You can change your residence from one neighborhood to another without getting permission from the city or the neighborhoods involved (condominium association boards notwithstanding).

All this is true also in the virtual world of Mastodon:

>> Every Mastodon instance (neighborhood) has a code of conduct that you have to agree to before you join (move in). Violating the code of conduct can get you in trouble (a suspension or worse).

>> Regardless of whether your instance has a governing board, the administrator is definitely the permanent mayor, in charge of keeping things going.

>> While your instance administration can't tax you (yet) to keep the servers running, you do have an obligation to help out if you can. As the mother instance, `https://mastodon.social` helps other instances financially, but each independent instance has to keep the lights on.

>> Aside from a disciplinary situation, it's easy to change your Mastodon neighborhood — and bring all your friends with you.

TIP

>> If you ever need to move your account to another instance, check out the following for instructions: `https://docs.joinmastodon.org/user/moving`.

Chapter **2**

Digging into Mastodon

I n spite of what you may have heard, signing up for Mastodon is simple. The first step in joining Mastodon is to choose a server, also known as an *instance*, to be your home in the fediverse. In the same way that getting a phone number from any provider allows you to call anyone else with a phone number, setting up a Mastodon account on any Mastodon server gives you access to the entire fediverse.

In this chapter, you learn how to choose a server, and then you sign up for an account and customize your profile. Then you see how to set the preferences for your account, get verified, and secure your account.

Choosing a Server

Mastodon is a decentralized social media platform that's made up of thousands of separate computers (also known as *servers* or *instances*) that have all agreed to speak the same language and share certain data.

When you create your Mastodon account, you get a username, which will be tied to one of these servers. Each server in the network has its own personality — and some servers pride themselves in having no particular personality. Every server is free to join, although some may be closed to new accounts or may require that you get on a waiting list.

The most difficult part of getting a Mastodon account is choosing which server to join. Fortunately, the website at https://join mastodon.org, which is shown in Figure 2-1, can help you sort through your options.

FIGURE 2-1: Your first stop should be joinmastodon.org.

You can sign up for Mastodon using the app or the website. However, signing up on the website — on your desktop, laptop, tablet, or phone — is easier than signing up with the app.

In this chapter, we demonstrate signing up and customizing your account mostly using the website.

REMEMBER   Whichever method you choose, go to https://joinmastodon.org. Then click the Get the App link to use the app or click Create Account to use a web browser. Both the website and the app are free, and creating your account on one allows you to use all the features of both.

If you click Get the App, you'll be taken to a screen where you can choose to go to the Apple App Store (to download the iPhone app) or to Google Play (to download the Android app). Once you get to the app store, download and install the Mastodon app as you would with any app.

Signing into the Mastodon mobile app is covered in the "Signing in Using the Mobile App" section, later in this chapter.

Because Mastodon is based on a free and open standard, there are many apps to choose from to use Mastodon. Start with the official one, which is called Mastodon. Other apps that work with Mastodon are covered in Chapter 8.

Browsing your options

When you visit https://joinmastodon.org and click the Create Account button, you'll be taken to the Servers page, which is shown in Figure 2-2.

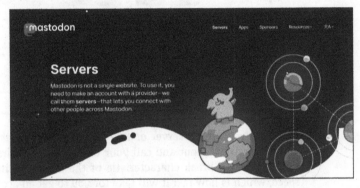

FIGURE 2-2: Checking out the servers.

Many more Mastodon servers than the ones you see when you scroll down this page are available. However, the servers on join-mastodon.org follow certain rules known as the Mastodon Server Covenant, which is an agreement by the server's owner to work to keep the server free of hate speech, to back up their data daily, to have a person on call to deal with technical issues with the server, and to give users of the server at least three months' notice before shutting down the server. Because these servers have agreed to the Server Covenant, they are seen as trustworthy, stable, and safe places to serve as your home in the Mastodon universe.

Although you can't go wrong with any of the instances listed on the servers page, some will be a better fit for you than others — and that's what makes Mastodon so cool!

Spend some time browsing through the different servers listed here. Note that you have several options for filtering the results — including by geographical region and topic (listed on the left side of the screen) and by legal structure, sign-up speed, and language (selected from drop-down lists at the top), as shown in Figure 2-3.

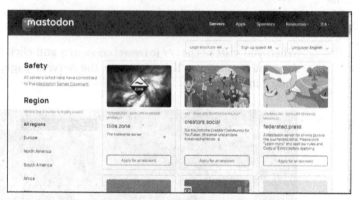

FIGURE 2-3: Reading the server descriptions.

Once you've found a server or two that sound like places where you'd like to hang out and call your home base, the first potentially make-or-break characteristic of the server is the *sign-up speed*, which is how fast it will take for you to get an account.

Understanding sign-up speed

When you browse the list of servers, you'll see two types of buttons: Apply for an Account button and Create Account. The Create Account button means that the server has instant signup: You can have a new account on that server in the amount of time it takes for you to choose a username and a password. Servers with an Apply for an Account button require that new accounts be approved by an administrator. This approval may take only a few minutes or a day. If you're super-excited to start using Mastodon, go with one of the ones that have the blue Create Account button.

TIP

Don't worry too much about picking the perfect server. Choosing a server doesn't have to be permanent. You can move your account (and all your followers!) to another server easily after you sign up.

Reading the rules

Regardless of whether you choose to apply for an account or go with a server that has instant account creation, check out the rules of your chosen server before you commit. To see a server's rules, first click the Create Account or Apply for an Account button to go to the server.

Once on the server, you'll see a description of the server on the left, with the name of the server administrator and the number of users. Below the server information, click Learn More to read more about the server, including its code of conduct, which describes the guidelines for how users of the server are expected to behave while using the server.

REMEMBER

Rather than having a complicated set of rules that try to make the greatest number of people happy and end up pleasing no one (such as what centralized platforms such as Twitter and Facebook must do), each Mastodon server sets specific rules and the conduct expected from users. If you don't like the code of conduct on one server, check out other servers until you find one with a code of conduct more in keeping with your beliefs.

Previewing a server

When you sign up using a web browser, many of the servers listed at joinmastodon.org allow you to browse the content on the server before joining. If you click to the server and see posts, rather immediately seeing a sign-up or login page, you've come to one of these. To see the content posted by the local users of the server, click the Local link on the right side of the page. The posts listed under Local are by people who would be your neighbors if you were to join this server. Are they talking about things you're interested in? If so, this may be the perfect server for you!

Meeting the admin

Another way to choose a server is to find out more about the person running it. Every server prominently displays the person in charge of keeping the server running and enforcing the server's rules. Click their profile from the server's home page or from the server's About page to read more about them and to see the things they're interested in and what they post.

On Mastodon, the admin sets the tone for the server. If the admin seems like someone you'd like to hang out with, that's a good indication that you've found a home.

Seeing why smaller is better

A server might have just a handful of members to many thousands. Servers with a smaller number of users are vital to keeping the entire universe interesting and avoiding a monoculture. Smaller servers also tend to have stronger personalities than larger servers. And smaller servers are often less prone to slowing down because they're less likely to get a lot of traffic.

Because every server allows access to the other servers, you have nothing to lose by going with a smaller one. In the same way that you're more likely to make friends in a small community group versus an international club with millions of members, joining a smaller Mastodon server makes you a bigger fish because the pond is smaller.

Signing Up

If you can't decide on a server, just pick any server that lists its topic as General. As mentioned, you're not stuck with your choice, and it's common to switch servers down the road once you get the hang of using Mastodon.

Follow these steps to begin to create your Mastodon account.

1. **Click the Create Account or Apply for an Account button on joinmastodon.org for the server you want to join.**

If you're using a web browser, you'll see posts on the server or the sign-up (login) page.

2. **If you don't see the login form, click Create Account on the right side of the site.**

3. **On the Create Account page (an example of which is shown in Figure 2-4), enter the requested information.**

This info will include your desired username, a password, and possibly your desired display name. More on these choices in the sections that follow.

FIGURE 2-4: Signing up on a server.

Choosing your username

WARNING

Your username on Mastodon can't be (easily) changed. Your username is the first part of your full Mastodon address, which resembles an email address in that it contains a username followed by the name of the server.

It's common to make your username some variation of your real name: for example, your first and last name without a space between them (chrisminnick), or with a period between them (chris.minnick), or with some random series of numbers after your name if you have a common name (chris.minnick153). However, your username can be anything you want — as long as the name hasn't already been taken by someone else on the server.

Choosing a display name

REMEMBER

Your display name, like your username, doesn't have to be your actual name, but there's nothing wrong with using your real name. Some servers enable you to choose a display name during the sign-up process, but others make you wait until after you're signed up.

Here are some tips for choosing your display name:

>> Your display name, unlike your username, can be changed at any time.

>> Your display name doesn't have to be unique to just you. If you want to use the same display name as someone else, that's possible but not recommended.

>> Display names can contain spaces.

>> You can use emojis in your display name! To see a list of the emojis supported by your server, visit https://emojos.in/ your server name. For example, if your server is mastodon. social, visit https://emojos.in/mastodon.social. Click any emoji to copy its code to your clipboard, and then paste it into Mastodon.

Note that the address to the emojis page starts with *emojos*. Emojos is a term used in Mastodon to refer to custom emojis.

Choosing a password

A password on any website is your first defense for keeping your account secure and making sure that no one can take over your account. Hacked accounts don't seem to be a problem on Mastodon currently, but hacked accounts will certainly increase as users increase.

Choose a password that would be difficult to guess. It's never a good idea to use your birthday or pet's name as a password. Try using the first letters from the first line of your favorite song to create a password that's both secure and random enough to be difficult for someone else to guess. Better yet, consider using a password generator such as the one built into your web browser or use a password manager (such as 1Password, which is available at https://1passowrd.com, or Bitwarden, which is available at https://bitwarden.com).

Once you've entered your password (twice) and clicked the button at the bottom of the form (which might say Sign Up or Get on the Waitlist or something else) you'll get a message that an email has been sent to you, as shown in Figure 2-5. When you get this email, click its link, and you either see a message thanking you for confirming your email address or are logged in.

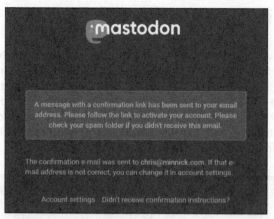

FIGURE 2-5: Congratulations! Click the link in the email, and then continue to the final step.

Following the suggested people

Because every server is slightly different, the next step may vary. However, your server will likely have a list of suggested accounts to follow, as shown in Figure 2-6. Your server admin (or admins) have chosen these accounts as great accounts to follow to help you get started with Mastodon. It's a good idea to follow some (and possibly all) of these right off the bat so you don't get into Mastodon and see nothing but blank timelines.

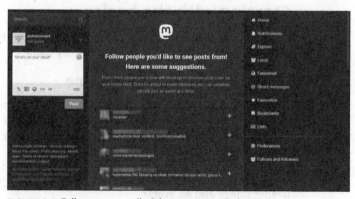

FIGURE 2-6: Follow some or all of the recommended users.

Customizing Your Profile

Now that you're signed up, the next step is to customize your *profile*, which controls how you look to the rest of the fediverse. To customize your profile, you use the Edit Profile screen. To access the Edit Profile screen, look for your username on the left side of the website interface (or click the icon in the lower-right corner of the app) and click the Edit Profile link (on the web) or the Edit Info link (in the mobile app), as shown in Figure 2-7.

FIGURE 2-7: Click Edit Profile under your username.

Starting with the basics

The most important piece of information on the Edit Profile screen, shown in Figure 2-8, is your display name.

Determining your display name

Depending on the server you chose to sign up with, you may have already created a display name when you created your account. If not, go ahead and do that now. For details, refer to the "Choosing a display name" section, shown previously.

Once you have a display name, you can enter a bio and upload a header image and an avatar.

FIGURE 2-8: Adding a display name, bio, header image, and avatar.

Writing your bio

As on other social media platforms, your bio is where you can tell other users of the platform a bit about yourself. When choosing what to reveal here, think about what you might want to have a conversation about with other people. Where you work? Where you live? Your hobbies and interests? If you're stumped as to what to write, see what other users have written.

One important consideration when writing your bio is that Mastodon doesn't have full text search — you can search only by users, URLs, and hashtags. Since you're new to Mastodon and people are unlikely to be searching for your username, you may want to include in your bio hashtags for subjects that interest you.

REMEMBER

Hashtags are words or combinations of words (without spaces) pre-fixed with the # symbol, such as #golf, #computers, #swimming, and #videogames.

Adding a header image

A *header image*, if you upload one, is shown in the rectangle at the top of your profile page. The header image will be displayed three times as wide as it is tall. It's best to crop your image to the right proportions before you upload it.

If you upload an image that isn't the right proportions, it will be cropped from the middle, so make sure that what's in the middle of the image is what you want to highlight on your profile page.

To add a header image, click Choose File under the Header section (to the right of the screen) and search for and select the file you want to use.

Add your avatar

Your *avatar* is the picture that appears with every one of your posts, as well as on your profile page. You might decide to have a picture of your face, or a cartoon version of yourself, or maybe just an image that you like. Since your avatar will generally be displayed only as a small image, it's a good idea to avoid using epic landscape photos with you in the distance.

The avatar will be displayed in a square, so choose a picture that's square or crop it to be a square.

To upload an image for your avatar, click Choose File under the Avatar section (to the right of the screen) and search for and select the file you want to use.

Setting options and metadata

Now that you've entered a display name, bio, avatar, and header, it's time to set some basic options. Before you do, however, click the Save Changes button at the top of the Edit Profile screen. You see a message that your changes have been saved. Now you can work on the next steps without worrying about losing all the hard work you put into your bio and uploading pictures if you accidentally close your browser window or the app.

Scroll down the screen a little and you'll see the check boxes shown in Figure 2-9.

Choosing follow settings

The first group of options controls how other users find and follow you. Here's a quick summary of when you might want to select or deselect each one:

>> **Require Follow Requests:** Select this box if you want to be followed by only users that you approve. In Chapter 5, you learn how to set who can read your posts. If you want to restrict your posts to being viewable only by your followers, you might also want to restrict who can follow you.

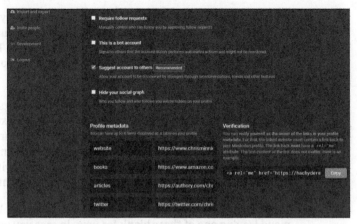

FIGURE 2-9: Choosing basic options and setting metadata.

>> **This Is a Bot Account:** Chances are good that you're not a bot (but do you know for sure?). If you are creating an account that will be posted to by a bot, select this box to tell other users that you're not a live person.

>> **Suggest Account to Others:** When this box is selected, Mastodon automatically recommends you to other users who might be interested in the topics you write about. You should select this box if you're interested in getting new followers.

>> **Hide Your Social Graph:** Your *social graph* is information about who you follow and who follows you. By default, this information is viewable to everyone on Mastodon. If you don't want other users knowing who you follow or who follows you, select this check box.

Entering profile metadata

Metadata is a fancy term for information that provides information about other information. For example, when you take a picture with your phone, the phone automatically stores information about the time and maybe even the location where the picture was taken. This information (the time and location) is metadata, or information about other information (the picture).

Your profile metadata is intended to give other users more information about your profile. Although it's not required, you can use profile metadata to create links to your profile on other websites or to websites you own. For example, you can link to your profile on Facebook or to your LinkedIn account to give more information about yourself than will fit on your Mastodon profile. These links will be especially important when you verify your account, which you'll learn about later in this chapter.

You enter profile metadata using the table that appears under the Profile Metadata heading (refer to Figure 2-9). The text you enter in the left column of each row of metadata will appear to the left of the text you enter on the right of each row. For example, if you want to link to your Twitter profile, you can type **Twitter** on the left and the address to your Twitter profile on the right.

You can enter up to four items in your metadata table. We recommend leaving one of these blank for now, in case you want to verify your account (as described in the "Getting Verified" section, later in the chapter).

Viewing and tweaking your profile

Once you've finished filling out your profile metadata, you're done with the Edit Profile screen (for now). Click the Save Changes button at the bottom of the screen (or the one at the top). After the message appears telling you that your changes were saved, click your avatar on the page to view your profile. Your profile will have the same basic layout (but not the same information, we hope!) as the one shown in Figure 2-10.

FIGURE 2-10: Viewing your profile.

You can view any user's profile (including your own) by clicking their avatar anywhere you see it in Mastodon.

Check your spelling and other information on your profile. You can go back to the Edit Profile screen to make changes. (Don't forget to click Save Changes.) When you're happy with your profile, move on to the next section!

Choosing featured hashtags

We can't stress enough how important hashtags are in Mastodon. To make it even easier for you to organize your content by hashtag and to make it easy for other users to find your content, you can set featured hashtags.

Featured hashtags should be hashtags that you use (or intend to use) in your posts. For example, if you frequently post pictures of your cat, you might want to use #catpictures as a featured hashtag.

Not every server offers the featured hashtags feature. If yours doesn't, it's not a big deal and you have one less thing to configure! Or you can send a message to the server's administrator and ask whether they plan to offer it in the future.

To set your featured hashtags, go to the Edit Profile page by clicking the Edit Profile link (on the web) or the Edit Info link (in the mobile app). Then click the Featured Hashtags link to display the screen shown in Figure 2-11.

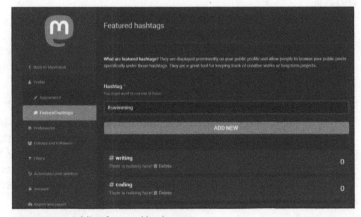

FIGURE 2-11: Adding featured hashtags.

To add a featured hashtag, enter the hashtag (including the #) in the text box and click Add New. Your hashtag will be added to your list of featured hashtags, and the text box will be ready for you to enter another one. You can add as many or as few hashtags as you like.

Featured hashtags appear on your profile with the number of times you've used each. The hashtags are clickable, so other users can see everything you've posted about that hashtag.

Setting Your Preferences

Preferences are settings you can change to improve or customize your experience using the Mastodon website or app. In this section, we talk about some of the more important preferences.

To get to the Preferences page, do one of the following:

>> On the web, choose Preferences from the three dots menu next to your avatar and username or from anywhere you see the Preferences menu item on the left or right of the screen.

>> In the mobile app, click the gear icon in the upper-right corner.

Some of the preferences that you can set on the website are different from those you can set in the app. For this section, we mostly describe features available on the website.

When you go to the Preferences screen, the Appearance preferences will be selected and highlighted in the left menu by default. These preferences control some essential aspects of how Mastodon will work and how (and whether) certain features will appear.

Choosing your preferred language

Mastodon has been translated into many different languages. If you would like to use a different language than the default language of your server, select it from the Interface Language drop-down list. The Interface Language preference controls the language displayed for buttons, help, and labels in Mastodon.

Checking out the themes

The Site Theme setting controls the colors and fonts you see. If you prefer a dark background with light text, select the Mastodon (Dark) theme. If you find dark text on a light background to be easier on the eyes, select the Mastodon (Light) theme. If you want a really dark background with really light text, choose the Mastodon (High Contrast) theme. Figure 2-12 shows the light theme, which is the one we'll use going forward to save ink.

FIGURE 2-12: Selecting a theme.

Trying out the advanced web interface

If you've ever used the Tweetdeck feature of Twitter, you'll understand and appreciate Mastodon's advanced web interface. The *advanced web interface* displays different types of data in multiple thin columns in your web browser. This interface helps you track and monitor multiple timelines without switching between them and makes full use of a large computer monitor.

While you're just getting started with Mastodon, we recommend sticking with the default interface. Once you get the hang of that, you may find that the advanced web interface is how you prefer to use Mastodon, or you may better appreciate the simplicity of the single-timeline view and decide to stick with that.

Looking at animations and accessibility preferences

Next up are the animations and accessibility preferences. These settings control things like whether animations (such as those in memes) automatically play, whether your feed automatically scrolls (which can be difficult to keep up with if you follow a large number of accounts) and whether to use system fonts (which can make reading posts on Mastodon easier for some users). Unless you know you want to enable or disable these preferences, our advice is to leave them as they are for now and revisit them later.

Glancing at the other options

If you scroll down the list of the other preferences, you'll see that most are self-explanatory. The options at the bottom that relate to sensitive content, however, warrant some further explanation.

When you post on Mastodon, you can choose to mark your post as sensitive. For example, if an image might upset other users, you can give a warning and your posts will be hidden behind the warning unless a user specifically clicks the warning.

If you want to see everything or you want to always decide whether you want to see something, change the Media Display preference to Always Show or Always Hide, respectively.

The last option on the page pertains to content warnings. Content warnings are similar in purpose to sensitive media, in that they let posters indicate that their post may not be for everyone. If you want to always see content that users have marked with content warnings, rather than having to click to view it, select the Always Expand Posts Marked with Content Warnings check box.

Getting Verified

Once upon a time, *verification* was a tool introduced by Twitter to let people know that an account of public interest was authentic. The blue check mark Twitter used to mark a user as verified became a status symbol. At least, that was the case until Twitter under Elon Musk opened up verification to any user with $8 a month to spend — an idea that quickly proved to be flawed as users became verified and impersonated other verified accounts.

Mastodon doesn't have an official verification process. However, Mastodon does give you the ability to verify that a link you list in your profile belongs to you. Verified links are listed with check marks on your profile. By verifying a link to an official link (such as your personal blog or another social media site), you can gain the benefit of proving that you are who you say you are without having to spend $8 per month or having to be a public figure.

The one catch is that you need to paste the link into a site where you have access to the code. Simply posting your verification link to your Twitter or Facebook profile won't work.

Getting your verification link

To verify a link, you need to put a special link to your profile on a site you control. Begin by visiting your Edit Profile screen and scrolling to the bottom, as shown in Figure 2-13.

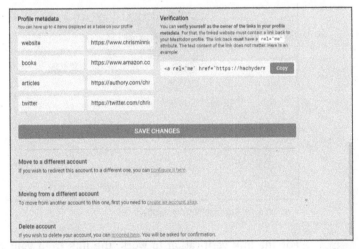

FIGURE 2-13: Getting your verification link.

To the right of the Verification link text box, click Copy to copy the link to your clipboard. This link contains a special rel="me" code that Mastodon requires to verify your link.

Once you've copied your verification link, you need to install it on a site you own and to which you have access to the code. If you have a blog site (such as a WordPress site or a Wix site), you can paste the verification link into the header or footer of your site's home page template.

After you post the verification page to your website, visit the page where the link appears and copy the page's address from your browser address bar. You'll need that address for the next step of the verification process.

Linking back to Mastodon

Once you've posted your verification link to a website you own and copied the exact address for that page, return to your Mastodon account and open the Edit Profile page.

Paste the link you copied in the preceding section into one of the Profile Metatag text fields (refer to Figure 2-9) and click Save Changes. If you did everything correctly, you'll see a green background around the verified link along with a check mark to the left of it, as shown in Figure 2-14.

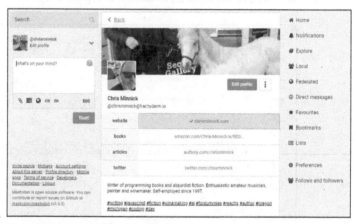

FIGURE 2-14: Success! You're verified.

WARNING

For the link verification to work, the link from your Mastodon profile must be exactly the same as the address that shows in your browser's address bar when you visit the page you're verifying. Also, verification can sometimes take a few minutes or longer, so you might need to wait anywhere from a few minutes to a few hours to see whether it worked.

But can I have a blue check mark?

Mastodon doesn't have an official way to mark your username with a verification mark. However, if you must have a blue check mark, many servers allow you to use emojis in your display name (as described in the "Choosing a display name" section, earlier in this chapter).

If your server allows the use of a verified emoji (which can be inserted using the :verified: code), you're free to add that to your display name at any time. Or why not be creative and choose an alternative icon to indicate that you're not interested in any status symbol? For example, you might choose the :verified paw: emoji, shown in Figure 2-15, to indicate that your pets have verified that you are who you say you are.

Chris Minnick 🐾
@chrisminnick@hachyderm.io

FIGURE 2-15: Check marks are used on that other site.

To add an emoji to your display name, just type the code for the emoji (starting with : and ending with :) at the end of your display name on the Edit Profile screen.

Securing Your Account

Although two-factor authentication (2FA) is not required, it's a good idea to enable it on any website where it's an option. Enabling *two-factor authentication* will require anyone who tries to log into your account (only you, we hope) to verify their identity.

Two-factor authentication makes it much more difficult for anyone to steal your Mastodon account and impersonate you. Logging in with two-factor authentication will be only an occasional hassle, much like any app or website that requires you to click a link or enter a code that they text to you to prove your identity.

Getting an authenticator app

To use two-factor authentication, you need to use an authenticator app on your mobile phone, regardless of whether you're using the app or the website on a computer or a mobile device. The most common authenticator app — and the one we use in this section — is Google Authenticator. You can download and install Google Authenticator from the Apple App Store or Google Play.

Setting up two-factor authentication

Once you have Google Authenticator installed and configured, go back to Mastodon and do the following:

1. **Click the Edit Profile link under your avatar.**

2. **On the Edit profile screen, click the Account link in the left navigation.**

   You see the two-factor authentication page shown in Figure 2-16.

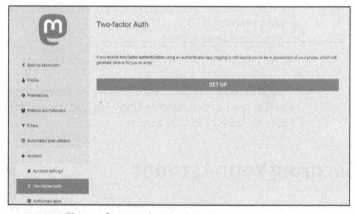

FIGURE 2-16: The two-factor authentication page.

3. **Click Set Up.**

   A QR code appears.

4. **Use your authenticator app to scan the QR code.**

   A new 6-digit code appears next to the name of your server.

5. **Click the 6-digit code to copy it.**

6. **In the text field on the Mastodon website or in the app, paste or enter the code that shows up in your authenticator app.**

   You see a list of recovery codes. In Figure 2-17, we've obscured the codes; you'll see a list of random characters.

REMEMBER

You'll need these codes if you lose access to the phone you used to set up two-factor authentication. Print this page or save the codes somewhere safe and never share them with anyone.

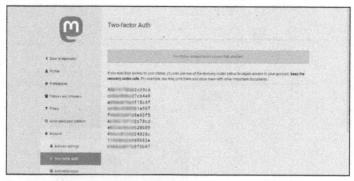

FIGURE 2-17: Keep your recovery codes safe and secret.

Understanding Mastodon's Three Timelines

Finally, it's time to start using Mastodon to converse with your fellow inhabitants of the fediverse. Unlike Twitter, which has only one timeline, Mastodon has three:

>> Home

>> Local

>> Federated

To make the most of your experience, you should understand the differences between them.

The home timeline

The *home timeline*, shown in Figure 2-18, is the default timeline you see when you first log in to Mastodon. You can get to the home timeline at any time by clicking the Mastodon logo or Home on the right of the screen.

This timeline displays posts only by people you follow. (Well, you may see posts written by people you don't follow if the posts are shared by someone you do follow.) The more people you follow, the more activity you'll see here.

FIGURE 2-18: The home timeline shows posts from people you follow.

The local timeline

To access the local timeline, click its link (below the link for the Explore timeline on the right). The *local timeline*, shown in Figure 2-19, displays the most recent posts by users of your home server. This is a great place to find out more about your closest fediverse neighbors and to get news about what's happening in your Mastodon community.

FIGURE 2-19: The local timeline displays recent posts by those on your home server.

The federated timeline

The *federated timeline* displays all the posts happening from every Mastodon server that your home server can connect to. The flow of posts in the federated timeline, shown in Figure 2-20, can be lightning fast and hard to keep up with.

FIGURE 2-20: The federated timeline displays all posts.

If you want time to read and interact with posts in the federated timeline, scroll down the screen. New posts will continue to accumulate at the top, but the posts at the bottom don't scroll by you. Alternatively, let the posts in the federated timeline fly by and gaze in wonder and confusion at the amazing amount of activity and diversity that makes up the entire world-wide distributed network.

Signing in Using the Mobile App

If you prefer to use the Mastodon mobile app, you can download it from the Apple App store or from Google Play. Figure 2-21 shows the app in Apple's App store.

After you download the Mastodon app, open it by tapping its icon from your phone's home screen. You'll see a screen similar to the one shown in Figure 2-22.

FIGURE 2-21: The Mastodon app in the App Store.

FIGURE 2-22: The Mastodon app's home screen.

Follow these steps to sign in with the account you created using the website:

1. **Tap the Log In button.**

   A screen asks for the name of your server. This is the same as the last part of your Mastodon address, as shown in Figure 2-23.

2. **Enter your server name and then tap Next.**

   You see the pop-up message shown in Figure 2-24.

3. **Tap Continue.**

   A screen appears, asking for authorization for the app to use information from your Mastodon account.

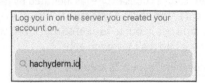

FIGURE 2-23: Entering the name of your server.

FIGURE 2-24: Allowing the app to sign in using your server.

4. **Tap the Authorize button.**

5. **If you're not already signed into your Mastodon server on your mobile device, enter the email address and password you used to create your Mastodon account and then tap Log In (see Figure 2-25).**

6. **If you have two-factor authentication enabled for your account, enter your two-factor code and then tap Log In (see Figure 2-26).**

   The steps for enabling two-factor authentication are described in the "Securing Your Account" section in this chapter.

7. **Open your authenticator app.**

   We used the Google Authenticator app, as shown in Figure 2-27.

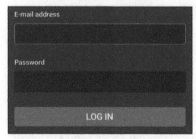

FIGURE 2-25: Logging in to your Mastodon account on your mobile device.

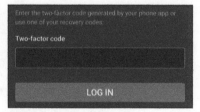

FIGURE 2-26: The two-factor authentication screen at login.

8. **Tap the numbers below your server name in Google Authenticator to copy them, and then return to the Mastodon app to paste the code into the text field.**

WARNING

Google authenticator's codes expire (as indicated by the pie chart to the right of the code). Make sure to copy the code when it's new (and not about to expire) so that it will work when you enter it into the Mastodon app.

The screen shown in Figure 2-28 appears.

FIGURE 2-27: Google Authenticator.

FIGURE 2-28: Welcome to the Mastodon app!

9. **Tap Find People to Follow.**

You see your server's list of recommended people to follow, as shown in Figure 2-29.

10. **If you want, select some people to follow and then tap Done in the upper-right corner of the screen.**

You see your home timeline in the app, as shown in Figure 2-30.

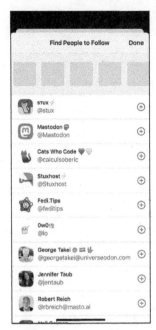

FIGURE 2-29: The Find People to Follow screen.

FIGURE 2-30: The Home timeline in the Mastodon app.

IN THIS CHAPTER

» Finding people by usernames and hashtags

» Following people

» Following hashtags

» Finding out what's trending

» Blocking users and servers

Chapter **3**

Finding and Following Friends and Strangers

Your home feed on Mastodon is made up solely of posts by people you follow. Without algorithms, advertisements, and sponsored posts interfering, you're free to follow as many (or as few) people as you like without having to worry about whether your choice of what you follow and read will influence what you see in the future. You'll also quickly notice that, unlike on Twitter, the number of followers you have versus the number of people who follow you has no bearing on whether you'll show up on other people's timelines.

While this new freedom to follow at will is liberating, you have to know how to find people you want to follow. In this chapter, you find out about some of the tools available on Mastodon for finding people and topics to follow. You also learn about the flipside of following: blocking and filtering people and servers you don't want to hear from.

Searching for Users and Hashtags

The search box in the upper-left corner of the Mastodon website or the mobile app can be used like you use any search feature on any other website. However, the basic search searches for posts and users only based on usernames and hashtags.

TECHNICAL STUFF

Mastodon's limited search feature was created this way by design. Full text search is often abused on other social media platforms to find people to *dogpile* — a form of online abuse in which groups of harassers target the same victim.

A form of full-text searching can be installed and enabled by the administrator of the server if they choose. If it's available on your home server, you'll be able to find the full text of your own posts, your favorite posts, your bookmarks, and posts that mention you.

To find out whether your server supports basic or full-text searching, click the search box. The help box shown in Figure 3-1 appears, telling you which options are available to you for searching.

FIGURE 3-1: The Mastodon search box and search help box.

Searching for users

If you know a user's Mastodon address, display name, or username, you can enter it in the search box to find that user. When you press Enter on your keyboard, the user's avatar, display name, and Mastodon address are displayed in the middle of the screen, as shown in Figure 3-2.

Click the follow icon to follow the person

Click here to see the person's profile

FIGURE 3-2: Searching for a user by display name, username, or address.

If you know you want to follow a user account shown in the search results, click the follow icon to the right of the search result.

Clicking a user shown in the search results will bring up the user's Mastodon profile, where you can find out more about them and read any posts they've made publicly available.

Searching for hashtags

Searching by hashtag is the best way to find content about specific topics. To search by hashtag, enter the hashtag (including the # symbol) in the search form and submit it. If you enter a hashtag that matches the beginning of other hashtags, those will be shown as well.

For example, searching for #cat returns hundreds of hashtags that start with #cat, as shown in Figure 3-3.

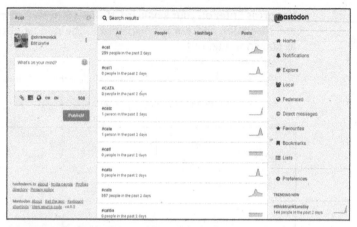

FIGURE 3-3: Finding hashtags with partial matches.

Following Other Users

On Mastodon, following another user adds their posts to your home timeline. It's as simple as that. Following a user doesn't associate you with them or influence the topics that show up in your app or when you're on the website. It most certainly doesn't influence which ads you see, because there are no ads. If you follow someone, they'll be notified that you followed them, and they may even look at your profile and decide they want to follow you back. And just like that, you've made a connection.

Follow freely

The more accounts you follow, the more interesting your feed will be. If you follow only a few users, your home timeline will be slow moving. This is different from Twitter, where your feed appears to be a place of constant action and drama. This is by design because the way Twitter makes money is by keeping you interested in being on Twitter for as long as possible.

On Mastodon, there are no tricks to get you to stay around longer. What you follow is what you get. So to get the most out of Mastodon, follow as many people as you want and don't be afraid to step outside your comfort zone and follow accounts that you might not follow on a commercial app.

Finding recommendation lists

One great way to find interesting people to follow is by using recommendation lists. Many users of Mastodon have put together lists to help others gain an audience and to help you find people to follow.

To get started with following lists, search for @FediFollows on Mastodon and follow them. The @FediFollows@mastodon. online account is maintained by the creator of https://fedi. directory/, a curated list of accounts to follow on Mastodon.

@FediFollows regularly posts links to accounts that are worth following. Adding them to your feed is a great way to get a constant stream of new ideas for accounts to follow. Be sure to check out https://fedi.directory/ while you're at it.

Another directory of accounts to follow is Trunk for the fediverse at https://communitywiki.org/trunk. The Trunk website, shown in Figure 3-4, might not be the most advanced web user interface, but it's a great resource for finding lists of users who post on topics that might be of interest to you.

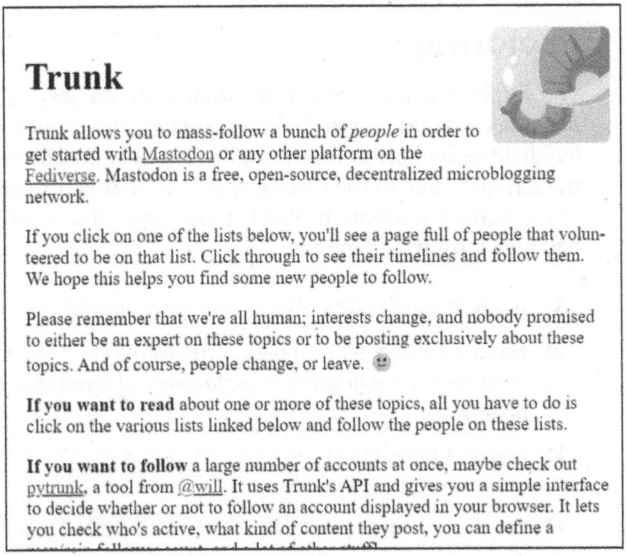

Trunk

Trunk allows you to mass-follow a bunch of *people* in order to get started with Mastodon or any other platform on the Fediverse. Mastodon is a free, open-source, decentralized microblogging network.

If you click on one of the lists below, you'll see a page full of people that volunteered to be on that list. Click through to see their timelines and follow them. We hope this helps you find some new people to follow.

Please remember that we're all human; interests change, and nobody promised to either be an expert on these topics or to be posting exclusively about these topics. And of course, people change, or leave. 😊

If you want to read about one or more of these topics, all you have to do is click on the various lists linked below and follow the people on these lists.

If you want to follow a large number of accounts at once, maybe check out pytrunk, a tool from @will. It uses Trunk's API and gives you a simple interface to decide whether or not to follow an account displayed in your browser. It lets you check who's active, what kind of content they post, you can define a

FIGURE 3-4: Trunk for the fediverse.

To use Trunk, scroll down the page to the list of topics. When you find a topic you're interested in, click the topic name. You'll see a list of users who write about that topic. Click a user to go to their profile, where you can read more about them and see their posts and then follow them if you want.

An even better feature of Trunk is that once you've started posting on Mastodon and have at least 20 posts, you can ask to be listed in the directory.

Following accounts boosted by friends

Once you have some accounts that you follow, you'll discover that if an account you follow boosts a post, that post will also show up on your timeline.

TIP

Boosts are Mastodon's equivalent of a Twitter retweet (reposting a post from someone else). Boosting posts is covered in Chapter 5.

If a post was boosted by someone you follow, there's a good chance that the boosted post was written by someone you might be interested in following. Click to their profile to learn more about them and follow them if you want!

Exploring

The Explore page is where Mastodon shows you accounts and hashtags that you might be interested in. Open the Explore page by clicking its button on the right side of the website. To access the Explore page in the mobile app, tap the magnifying glass icon in the bottom navigation. The Explore page, shown in Figure 3-5, has four tabs across the top:

>> **Posts:** Popular posts from across the fediverse.

>> **Hashtags:** Popular hashtags, along with a chart showing whether the hashtag is currently trending upward or downward.

>> **News:** A news feed of articles that are being talked about in the fediverse right now.

>> **For You:** Accounts that you might be interested in following.

FIGURE 3-5: The Explore page.

The recommendations on the For You tab are generally users who have similar interests to yours. If you selected the Suggest Account to Others option in your profile settings, you'll appear in the For You tab of other users who have similar interests to yours.

REMEMBER

The Suggest Account to Others option is covered in the section on setting options and metadata in Chapter 2.

Following Hashtags

Whether you find a hashtag by using the search feature, by clicking it from a profile, or from looking at the Explore page, clicking a hashtag will display a list of posts from the entire Mastodon universe that use that hashtag.

If you frequently search for a certain hashtag, you might want to follow it. When you follow a hashtag, all the posts containing that hashtag will appear on your home timeline.

To follow a hashtag, go to the hashtag's page by clicking it from anywhere in Mastodon. On the hashtag's page, look for the follow icon in the upper-right corner of the screen. The follow icon on a hashtag page is the same follow icon you'll see next to user account info — a person with a plus sign, as shown in Figure 3-6.

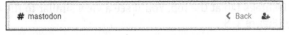

FIGURE 3-6: The follow icon for following a hashtag.

Clicking the follow icon changes it to an unfollow icon, which is a person with an x. If you followed a user or a hashtag, you can unfollow them by clicking this unfollow icon.

Viewing Trending Topics

As with any social media site, popular topics come and go on Mastodon. Somehow, Mastodon's trending topics seem to be more random than the usual sports or current events trends. On #thicktrunktuesday, pictures of large trees are all the rage. November 23 is #wolfenoot, a day for celebrating people who are kind to dogs. Hashtags in languages other than English often make it into the list of trending topics, and a person can learn a lot by seeking to understand what's causing all the interest.

No matter what's trending, a great way to find out about your fellow Mastodon users is to check out what goes viral.

Viewing top hashtags

To see the current top hashtags, check out the Trending Now graphs in the lower right of the website (if the feature is enabled by you and your server's admin). The Trending Now area is constantly updated and shows just a few of the most popular topics being talked about on the network at the moment, as shown in Figure 3-7.

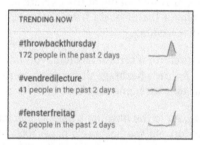

FIGURE 3-7: The Trending Now graphs.

Want to see more hot hashtags? Visit the Hashtags tab in the Explore view. Click Explore on the right of the website, and then click the Hashtags tab at the top. You'll see a view similar to the one shown in Figure 3-8.

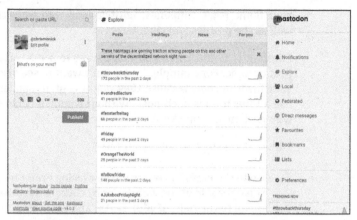

FIGURE 3-8: Viewing more trending hashtags on the Hashtags screen.

Click any of the hashtags in the Hashtags window to see posts that use that hashtag.

Learning from history

History often repeats itself, and some trending topics come back week after week on Mastodon. A popular hashtag that trends every week is #followfriday. On #followfriday, many people post accounts that they think are worth following. If one of your friends or someone you follow participates in #followfriday, this is a great opportunity to find more people to follow. If you're featured in someone else's #followfriday post, follow people back who follow you as a result.

Blocking and Filtering

It's not always sunshine and roses on Mastodon. There is content you might not care to see, users you might not want to see, and even entire bad servers. Mastodon gives both the server's administrator and the users of the server the power to block and filter users and posts they don't want to see.

Filtering content

Mastodon's filter capability allows you to specify keywords or phrases and hide any content containing those keywords from your home page, timelines, notifications, and conversations, as well as the profiles you visit.

Filtering is useful when you don't want to see a certain kind of post or topic. For example, if you don't want to see discussions of your favorite show because they might contain spoilers, you can use a filter to hide such posts without completely blocking users who post about that show.

Follow these steps to set up a filter:

1. **Go to the Preferences screen by clicking Preferences on the right side of the website or by clicking the three-dot menu to the right of your avatar and username.**

TECHNICAL
STUFF

The icon with three dots stacked vertically or horizontally is sometimes called a kebab menu because it resembles the delicious skewered dish originating from Middle East cuisines.

2. **Click Filters on the Preferences page.**

The Filters page appears. If you don't have any filters, you'll see just the Add New Filter button.

TIP

If you're using the mobile app, tap the gear icon (in the upper-right corner) to open Preferences, and then scroll down the page and tap Account Settings. On the next page, tap the three-line menu (sometimes called the hamburger menu) in the upper right and select Filters.

3. **Click the Add New Filter button.**

The Add New Filter form appears, as shown in Figure 3-9.

4. **Give your new filter a name and, if you like, select an Expire After time.**

5. **In the Filter Contexts area, choose in which areas you want to hide content that matches the filter.**

If you want to hide only filtered content from your home timeline and lists you create, for example, you can select the Home and Lists check box under Filter Contexts. To hide the content everywhere on Mastodon, select all the check boxes.

6. **Select whether you want to hide the matching content or put it behind a warning.**

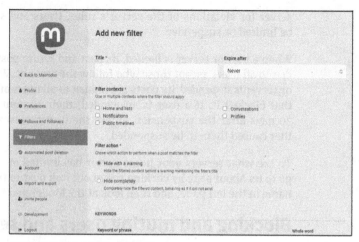

FIGURE 3-9: The Add New Filter form.

7. **Add one or more keywords or phrases you want to filter out, as shown in Figure 3-10.**

8. **Click Save New Filter.**

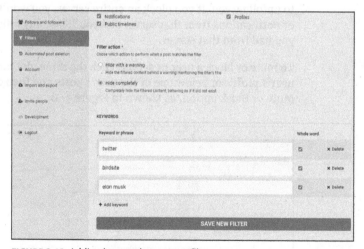

FIGURE 3-10: Adding keywords to a new filter.

Finding out who's blocked

On Mastodon, blocking is performed by the server administrator or a *moderator*, who is a person given the power to monitor the server for violations of the server's rules. Users and servers can be limited or suspended.

When a user or server is *limited*, its past and future posts are hidden to all users except those who follow the account. When a user or server is *suspended*, its posts are hidden to all users of the server that blocked it. If a user is suspended, their account is deleted 30 days after the suspension unless the user resolves the issue that caused them to be suspended.

To see what servers your home server has limited or suspended, go to its About page by clicking the About link next to the server's name in the left pane, and then look at the Moderated Servers list.

Blocking and muting users and servers

You have the power to block users or entire servers from showing up on your account. To hide all content by a user or server, you have two options: muting and blocking. *Muting* prevents posts from a user or server from showing up in your timeline without the muted user knowing you've muted them. *Blocking* prevents a user's posts from showing up in your feeds but also forces them to unfollow you. If you block an entire server, you won't see posts or notifications from that server and you'll lose any followers that you had from that server.

To mute or block a user or server, click the three-dot menu on the user's profile or below one of the user's posts and then choose the Mute or Block option, as shown in Figure 3-11.

FIGURE 3-11: Blocking or muting a user.

When you click the link to mute a user, a confirmation message appears, asking whether you're sure and how long you want the mute to last, as shown in Figure 3-12.

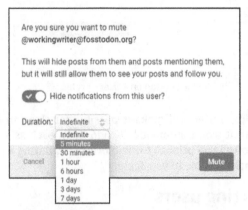

Are you sure you want to mute
@workingwriter@fosstodon.org?

This will hide posts from them and posts mentioning them, but it will still allow them to see your posts and follow you.

Hide notifications from this user?

Duration: Indefinite

Indefinite
5 minutes
30 minutes
1 hour
6 hours
1 day
3 days
7 days

Cancel   Mute

FIGURE 3-12: Are you sure you want to mute them?

When you block a user, you get a confirmation message asking whether you'd like to also report the user. When you report a user, a message is sent to the server's administrator and moderator.

When you block a server (by selecting Block Domain from the three-dot menu), you get a pop-up message asking whether you're sure. You also see a reminder that blocking an entire server removes your followers from that server and you'll no longer see any content from it.

WARNING

In most cases, it's better to block or mute individual users who you don't want to see rather than entire servers. However, if you notice that a large amount of content you don't want to see comes from a particular server, it's good to know that you have the power to stop seeing it.

Unmuting and unblocking

If you block or mute someone or a server by mistake, or if you change your mind about blocking or muting them, you can unblock or unmute them.

To unblock or unmute a user, go to their user profile. Click the Unblock or Unmute button, which appears where the Follow button usually is, as shown in Figure 3-13.

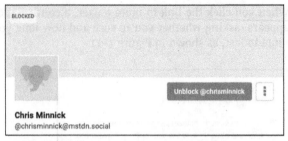

Chris Minnick
@chrisminnick@mstdn.social

FIGURE 3-13: The Unblock button on a user's profile.

Unblocking a user will make it possible for them to follow you again, but it won't automatically add them back as a follower if they followed you prior to being blocked. Unmuting a user makes their content visible to you again.

Reporting users

If you see content that you believe violates the server's rules, you can report the user who posted the content. To report a user or server, click the three-dot menu on their profile or a post and then choose the Report option.

When you report a user's profile or a post, a form appears so you can say why you're reporting them, as shown in Figure 3-14.

Reporting Test@mastodon.social

Tell us what's going on with this profile

Choose the best match

○ **I don't like it**
It is not something you want to see

○ **It's spam**
Malicious links, fake engagement, or repetitive replies

○ **It violates server rules**
You are aware that it breaks specific rules

● **It's something else**
The issue does not fit into other categories

Next

FIGURE 3-14: Specify why you're reporting a profile or a post.

Chapter **4**

Developing a Following

Mastodon doesn't have as many users as Twitter or Facebook (yet!), so developing a large following on Mastodon can be more difficult. However, because the Mastodon community is a small pond, it's also easier to be a big fish. In this chapter, you gain effective tools for starting to build your own following.

Finding Your Friends on Mastodon

As Mastodon has become more popular, millions of people have flocked to it from Twitter (or the *bird site*, as it's known on Mastodon). Perhaps you too first heard about Mastodon on Twitter. As a migrant from Twitter to Mastodon, you're in good company. Finding members of your Twitter flock on Mastodon isn't difficult, and it's a great way to get a head start with building a following.

Searching for fellow Twitter refugees

If someone considered you follow-worthy on Twitter, they will on Mastodon too. Several tools have been created for finding Mastodon accounts of your Twitter followers. If you haven't deleted your Twitter account, you can use any (or all) of these tools.

Using Twitodon

One service to help you find your Twitter followers on Mastodon is Twitodon, at https://twitodon.com. Twitodon works by linking to your Twitter account and your Mastodon account and comparing the list of accounts you follow on Twitter with accounts on Mastodon that have also used Twitodon. If matches are found, Twitodon will give you a file you can import into your Mastodon account.

Follow these steps to use Twitodon:

1. **Make sure you're logged into both your Twitter account and your Mastodon account.**

2. **Go to https://twitodon.com and click the link under the Get Started header to log into your Twitter account.**

   A window opens, asking you to authorize Twitodon to access your Twitter account.

3. **Click the button to authorize Twitodon.**

   You return to Twitodon and Step 1 has been crossed off.

4. **Enter your Mastodon server's address (including the https://) in the field under Step 2, as shown in Figure 4-1, and then click Sign In.**

   For example, if your Mastodon account is on mastodon.social, you'd enter https://mastodon.social.

   A screen appears asking you to authorize Twitodon to access your Mastodon account.

5. **Click Authorize.**

   You return to Twitodon and Step 2 has been crossed off.

6. **Watch the number of scanned users go up as Twitodon processes the list, as shown in Figure 4-2.**

   We hope the number of matches found go up as well!

REMEMBER

Because Twitodon can find only users who have also used Twitodon, don't be surprised if it doesn't come back with many or any results. The good news is that as more people migrate from Twitter and use Twitodon, it will be able to make more matches! So, it may be a good idea to try Twitodon again every so often as long as you keep your Twitter account.

they have logged into Twitodon previously. Because of this, we recommended you check back with us regularly to keep up-to-date as more people join the fediverse.

Information about how your data is handled is available in the Privacy Policy.

Get Started

Step 1. ~~Sign in with Twitter~~ Done. (Logged in as: @chrisminnick).

Logout and revoke access

Step 2. Sign in with Mastodon by entering your Mastodon host's web address:

Sign in

We have scanned 0 of 0 users you follow on Twitter and discovered 0 Twitter users on Mastodon who have previously linked their Twitter and Mastodon accounts.

Step 3. Download matching users in CSV format to import into your Mastodon account.

Step 4 (optional). Revoke our access to Twitter and Mastodon.

- Twitter: Use the button that appears when you are successfully logged into Twitodon with your Twitter account. Your login session has a time limit, so the buttons will not appear until you refresh your session by logging in again. When visible, click the Revoke button to disassociate Twitodon from your Twitter account.
- Mastodon: Unfortunately, Mastodon doesn't provide Twitodon the ability to revoke access, but you can still logout. To completely revoke access navigate to your Mastodon instance, log in, and go to your settings/preferences page, select "account" followed by "authorized apps", find the Twitodon app and click the revoke

FIGURE 4-1: Logging into your Mastodon server from Twitodon.

they have logged into Twitodon previously. Because of this, we recommended you check back with us regularly to keep up-to-date as more people join the fediverse.

Information about how your data is handled is available in the Privacy Policy.

Get Started

Step 1. ~~Sign in with Twitter~~ Done. (Logged in as: @chrisminnick).

Logout and revoke access

Step 2. ~~Sign in with Mastodon by entering your Mastodon host's web address.~~ Done.

(Logged in as: @chrisminnick@hachyderm.io)   Logout

We have scanned 372 of 968 users you follow on Twitter and discovered 5 Twitter users on Mastodon who have previously linked their Twitter and Mastodon accounts.

Step 3. Download matching users in CSV format to import into your Mastodon account.

Step 4 (optional). Revoke our access to Twitter and Mastodon.

- Twitter: Use the button that appears when you are successfully logged into Twitodon with your Twitter account. Your login session has a time limit, so the buttons will not appear until you refresh your session by logging in again. When visible, click the Revoke button to disassociate Twitodon from your Twitter account.
- Mastodon: Unfortunately, Mastodon doesn't provide Twitodon the ability to revoke access, but you can still logout. To completely revoke access navigate to your Mastodon instance, log in, and go to your settings/preferences page, select "account" followed by "authorized apps", find the Twitodon app and click the revoke button.

Step 5

FIGURE 4-2: Twitodon scans accounts you follow on Twitter.

7. When Twitodon finishes scanning, click the link in Twitodon's Step 3 to download a list of the matches that were found.

Twitodon's Step 4 (revoking Twitodon's authorization to access your Twitter followers) is optional, and we're going to skip it. Doing so will make it easier to come back and check Twitodon periodically in the future.

8. Click the link under Twitodon's Step 5 to go to the Import page on your Mastodon instance, or click Preferences, Import and Export, and then Import.

You see a screen similar to the one in Figure 4-3.

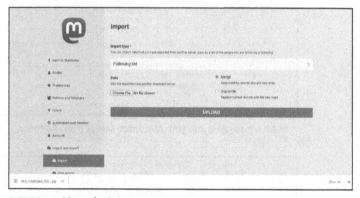

FIGURE 4-3: Mastodon's Import page.

9. Under Import Type, choose Following List (it should be selected by default).

10. Choose the Merge radio button (which should also be selected by default).

Merging means that you'll still follow everyone you follow on Mastodon before you import the matches found by Twitodon.

11. Click the Choose File button and locate the file you downloaded from Twitodon.

The file is named new_mastodon_follows.csv and should be in your Downloads folder.

12. Click the Upload button on Mastodon's Import page.

After a moment, you'll see a message that your file was uploaded and will be processed.

TECHNICAL STUFF

Depending on the size of the file you imported, it may take a few minutes or longer for the file to be processed and for the new users to be imported.

13. After a few minutes, go to your profile page and check whether your following number has gone up.

As you follow the people you followed on Twitter, they'll get notifications and some of them may decide to follow you back.

Using Fedifinder

Fedifinder (https://fedifinder.glitch.me/) can scan your entire Twitter account, including accounts you follow, accounts that follow you, and your Twitter lists to find Twitter users who list their fediverse address in their Twitter profiles. Because it scans your lists and followers and doesn't depend on matched users already having used it, Fedifinder is more likely to find results than other automated programs for finding other Mastodon users.

Follow these steps to use Fedifinder:

1. Make sure you're logged in to both Twitter and Mastodon.

2. Go to https://twitter.com/ and edit your Twitter profile to add your fediverse address in one of the following places:

- The description
- The location
- The website address

You can add your Mastodon address also to a pinned Tweet.

REMEMBER

Your fediverse address is your Mastodon server name followed by a slash, followed by your Mastodon username. It's the same as the link to your profile page on your instance. For example: hachyderm.io/@chrisminnick.

3. Go to https://fedifinder.glitch.me/ and click the Authorize Twitter button at the top of the screen.

4. Follow the instructions to authorize Fedifinder to access your Twitter data.

Fedifinder scans your Twitter account to find fediverse addresses and starts showing you the results, as shown in Figure 4-4.

Fedifinder
Join the Fediverse in 5 easy steps

1. Find an instance ✓
Your Twitter contacts are on these instances.

hachyderm.io mastodon 3.5.3, 20779 users, 5 of your contacts, **registration open**
indieweb.social mastodon 4.0.2, 9201 users, 4 of your contacts, **registration open**
infosec.exchange mastodon 4.0.1+glitch, 29841 users, 3 of your contacts, **registration open**
mastodon.cloud mastodon 3.4.6, 228786 users, 2 of your contacts, **registration open**
universeodon.com mastodon 4.0.2, 45341 users, 2 of your contacts, **registration open**
toad.social mastodon 4.0.3, 1674 users, 1 of your contacts, **registration open**

Show more

2. Publish your handle ✓
These handles were found in your profile @chrisminnick.

@chrisminnick@hachyderm.io

3. Scan your network ✓

FIGURE 4-4: Fedifinder scans your Twitter contacts for Mastodon addresses.

5. **Scroll down the page and click the link next to any Twitter lists you follow, and then click the Scan Followers link.**

6. **After all the scans have finished, click the Export CSV with Found Handles link.**

CSV, short for comma-separated values, is a way of storing lists of data in text files so they can be imported into other systems.

TECHNICAL
STUFF

7. **Go to the Import page on your Mastodon instance (click Preferences, Import and Export, Import) and import the resulting file (fedifinder_account.csv) in the same way you imported the file from Twitodon.**

Refer to Steps 9 to 13 in the preceding section.

Inviting friends and family

If you have a fairly large Twitter following and follow a large number of Twitter accounts, you may have acquired some new followers and follows from using Twitodon and Fedifinder. But not everyone you want to follow is on both Mastodon and Twitter. To get more people to join you on Mastodon, you need to invite them.

Mastodon makes inviting people to join your Mastodon server easy. Plus, when you use Mastodon's invite feature to invite people, signing up will be easier for them than it was for you because the link you give them will take them directly to the same local server you use.

If your server allows users to create invitations, you can get an invitation link by clicking Preferences and then Invite People. You'll see the Invite People page, which is shown in Figure 4-5.

FIGURE 4-5: The Invite People page.

The Invite People page allows you to create individual links, limited-use links, and links that expire. Unless you want to have a limited number of followers or you want to be able to track who uses which links, the best way to use Invite People is with the following settings:

>> Set Max Number of Uses to No Limit

>> Set Expire After to Never

>> Select the Invite to Follow Your Account check box

Once you've configured these settings, click the Generate Invite Link button. A text box with an invitation address appears, as shown in Figure 4-6.

Tap or click the Copy button to copy this link. Here are a few ideas for how you can use your Invite link:

>> Paste it into a post on Facebook.

>> Put it in your profile on Twitter.

>> Paste it into your Instagram profile.

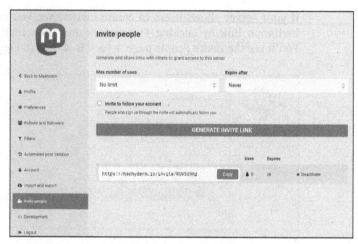

FIGURE 4-6: Your personal Mastodon invitation.

>> Email it to your friends and family and tell them to join you on Mastodon.

>> Send it in a text message.

>> Add it to a website you own.

When someone follows the link on their phone, tablet, or computer, they'll see the same sign up page you saw when you first signed up for Mastodon. Plus, when someone signs up using your link, you'll find out about it, the number of uses next to your invite link will go up, and you'll gain a follower!

Boosting and Favoriting

Each post on Mastodon has several icons below it for the different ways you can interact with the posts, as shown in Figure 4-7.

From left to right, the icons are as follows:

>> **Reply:** Opens the original post with a text input box under it where you can add a comment to the post.

>> **Boost:** Adds the post to your home timeline, where it will be visible to anyone who follows you. Boosting on Mastodon is like retweeting on Twitter.

Chris Minnick ⊙ ⊘ 5d
@chrisminnick@hachyderm.io

I'm working on a section for #mastodonfordummies about improving your
Mastodon experience.

Step 1: add some filters, like one to block references to Twitter. Here's a
screenshot from it. What other keywords should I add to this filter?

Reply   Boost   Favorite   Bookmark   Share

FIGURE 4-7: A Mastodon post, with the row of icons below it.

>> **Favorite:** Adds the post to your favorites list. Favoriting a
post (using the star icon) is like clicking the Like button on
Facebook. Posts that get a large number of favorites are
more likely to show up in the explore timeline.

>> **Bookmark:** Adds the post to your bookmarks list.

If you see a post you want to be able to find again later,
bookmark it, because content on Mastodon keeps streaming
and finding it later using the search feature is unlikely to
work.

TIP

>> **Share:** Displays the Share Link screen (shown in Figure 4-8),
where you can email a direct link to a post or copy a link to
the post. This feature is useful, for example, for posting
content from Mastodon to other social media sites.

A link you copy from the Share Link window can be followed by
anyone, regardless of whether they have a Mastodon address, as
long as the privacy settings of the post allow it. (Privacy settings
for posts are covered in Chapter 5.)

REMEMBER

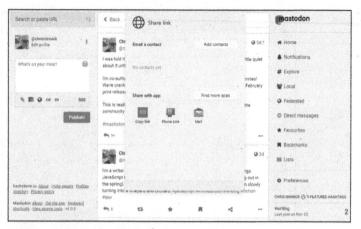

FIGURE 4-8: The Share Link window.

When you see content on Mastodon that you like, you can show your appreciation or agreement in several ways. Unlike on other social media platforms, content you click doesn't affect what you'll see in the future. And you won't suddenly see nothing but advertisements for cat food everywhere on the internet because you clicked someone's funny cat meme.

When you reply, boost, favorite, or share a post, the author of the post gets a notification. When other users interact with your posts, you also get a notification. It's common and encouraged on Mastodon to follow back people who follow you and who interact with your content. If you follow them, they are likely to follow you back.

REMEMBER

Getting large numbers of followers on Mastodon is not a competition, as it often seems to be on other sites. We can't stress this enough — follow people whose content you want to see when you log into Mastodon.

Joining in the Conversation

After you've been active on Mastodon for a few days, you'll start to notice some patterns that differ from those on Twitter. The biggest one is that people here seem to be friendlier, more welcoming, and less confrontational than on Twitter. In this section, you start to learn about how to interact with the community in ways that will help you get more interaction with your posts and more followers.

Introducing yourself

It's a tradition on Mastodon for new users to make one of their first posts an introduction. While there's no right way to post an introduction, most people follow some general guidelines.

The most important tip for writing your introduction is to add the hashtag #introduction to your post. People looking for new users on the site will be able to find your post using this hashtag, and it also serves as a sort of free pass to talk about yourself for a moment . . . maybe even in a marketing type of way (which is generally discouraged on Mastodon, except in introductions).

Keep your introduction short, and don't just repeat things that are in your bio. For the most engagement, talk about who you are and your interests. It's always important to use hashtags on Mastodon, but particularly important to do so in your #introduction. For example, here's an introduction that might be written by Leonardo da Vinci:

```
#introduction I'm Leonardo da Vinci. My friends
call me Leo. I do a bit of everything, and I love
learning. A few of my interests include
#painting, #astronomy, #botany, #cartography, and
#anatomy (I really love #dissection!),
#sculpture, #engineering, and #architecture.
#Florence #Milan #Rome
```

Asking for advice

Asking for advice from others (and using hashtags so the right people will see your post!) is a great way to build your community. Ask the right questions and you'll soon discover that Mastodon is filled with brilliant people who enjoy sharing their knowledge.

Asking big questions

Discussions of politics, economics, religion, sexuality, and the meaning of life are welcome on Mastodon. Server administrators seek to provide an environment where anyone can feel free to post their thoughts without fear of being harassed. Lively discussions of different interests and beliefs is part of what makes participating on Mastodon fun.

That said, if you post content that violates the terms of your Mastodon instance, expect to be quickly limited or suspended.

Being funny

It's not all serious discussion of important issues on Mastodon (not by a long shot). Puns, memes, videos, and pictures are all welcome and popular. Cat pictures are always popular, and if you log in on a Saturday (aka #caturday) it sometimes feels like half the new posts show adorable cats doing crazy things.

Being respectful of others

Harassing and trolling other users is seriously looked down upon on most respectable Mastodon instances. While some users and servers encourage trolling, they generally get quickly blocked by server administrators who seek to maintain a welcoming and friendly vibe.

Posting about Mastodon

With so many new users joining the fediverse and seeking to understand the best way to do things here, it's common to see questions and posts about Mastodon. If you don't understand a Mastodon feature, post about it. If you wish Mastodon had a particular feature, post about it. If you learn something about how Mastodon works that you wish had been explained to you when you first joined, post about it. If you fall in love with Mastodon and have a revelation that Mastodon is how social media was meant to be, post about that too!

TIP

Learning how to post and interact on Mastodon is a journey, and most of us are still figuring it out. In Chapter 5, you learn much more about how to write effective posts.

Using Direct Messages

When you want to talk to just one person on Mastodon, you can use a direct message. *Direct messages* are like normal posts but with the privacy setting set to direct so they are seen only by one person.

Writing and sending direct messages

To send someone a direct message, click the three-dot (kebab) menu on their profile or under one of their posts and select Direct Message @*username*, as shown in Figure 4-9.

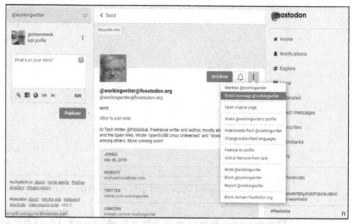

FIGURE 4-9: Opening a direct message post window.

A post window opens with the user's Mastodon address filled in. The address at the beginning of a direct message is the same as an email address. Type the content of your message after the user's address, as shown in Figure 4-10.

FIGURE 4-10: Composing a direct message.

After you send the direct message, it appears on your direct messages timeline, as shown in Figure 4-11.

FIGURE 4-11: Viewing your sent and received direct messages.

Remembering that direct is not private

Direct messages are not the same as the private messages you might send on other sites. They aren't encrypted in transmission between your server and their destination server, and they can theoretically be read by anyone who might intercept them. Your server administrator and the destination's server administrator can also read the contents of direct messages.

Because direct messages aren't encrypted, you shouldn't send any sensitive information (including passwords and secrets) in a direct message.

IN THIS CHAPTER

» Translating Mastodon slang

» Minding your manners on Mastodon

» Extending your reach with hashtags

» Posting multimedia content

» Enriching your posts

Chapter 5

Toot, Toot: Writing Effective Posts

n Chapter 4, we cover ways to find your friends and make new ones on Mastodon. In this chapter, we talk about what to say to them. You might have noticed that Mastodon posts have a different vibe than Twitter posts. In these pages, you learn more about the Mastodon mindset and how to participate effectively.

Knowing the Lingo

While you may be new to Mastodon, the platform and many of its residents aren't. The Mastodon community has developed its own jargon and ways of communicating.

Instances, nodes, and servers

Mastodon is a federation of independent sites running the same software. The software allows users of each site to communicate with each other, but each site makes its own rules about what kind of content is allowed.

A Mastodon site such as mastodon.social, fosstodon.org, or hachyderm.io is referred to as an *instance* of Mastodon or a *node* of the Mastodon network.

Mastodon itself is referred to as a *node* of the fediverse. Other fediverse nodes include Friendica, PixelFed, and PeerTube.

When people post about Mastodon *servers*, they're talking about the computers that contain the Mastodon software and the content moving through an instance.

Toots and posts

The origin story of *toot* sounds like an urban legend, too good to be true. But the story comes directly from Mastodon's founder, Eugen Rochko:

Probably not a lot of people know this now, but Mastodon's web app started out with "Publish". In 2016 a famous YouTuber jokingly offered to support Mastodon's Patreon forever if I changed the button to say "Toot" instead.

Needless to say, this was a really early and not very informed decision. The first glimpse of attention and financial support. As a non-native speaker I had no idea there was another meaning.

EUGEN ROCHKO, MAY 23, 2022 AT 12:25 PM

This decision led to a plethora of comments from users, some of whom decided not to use Mastodon. Others, however, were tickled by the term *toot* — one mobile Mastodon app is named Subway Tooter, and there are at least three Mastodon instances named Toot: toot.community, toot.io, and toot.works.

In the wake of the Great Mastodon Migration (when Elon Musk took over Twitter), the developers decided to change the Toot! button back to the Publish! button. Its fans, however, will likely continue to use the term *toot*.

TIP

You may have noticed that some of the screenshots in this book still display the Toot! button, while others use the Publish! button. As of this writing, the official Mastodon mobile app and some third-party mobile apps still use Toot! (and *reblog* instead of *boost*). We are confident that bug reports have been filed and expect that these terms will be consistent with the website in the next version of the apps.

Avoid using birdsite terms

Always remember where you are when you're posting. It's easy to fall into bad habits, especially after spending years on the birdsite, also known as Twitter.

Remember the following terms regarding posts:

> **Boost:** Use when sharing a post with followers. Don't use *retweet*.

> **Favourite:** Use when publicly favoring a post. (Note the UK spelling of *favorite*). Don't use *like (heart)*.

Content warnings (CW)

When scrolling through your feed, you may have come across a post or an image hidden behind a Show More button. If you clicked the button, you could view the post or image. Perhaps you wondered what was going on. What you found was a content warning (CW) from the poster.

Mastodon can be a wild, untamed, and not safe for work (NSFW) place. In addition, many communities that came together on Mastodon in the early days were victims of harassment on Twitter. Content warnings and marking images as sensitive allow users to choose whether to spend time with possibly upsetting content.

Some users put CWs on provocative political posts or items that might trigger post-traumatic stress issues. Others place a CW on posts dealing with controversial topics, seeking to avoid disrupting a peaceful vibe in people's feeds. CWs serve as a reminder that Mastodon should be a safe place for all kinds of people.

We talk a little bit about content warnings when you customize your profile in Chapter 2. Later in the chapter, we show you how to place a CW on your posts and mark images as sensitive.

REMEMBER

Whether or not to put a content warning on one of your posts is your decision. However, moderators can mark images as sensitive in response to a complaint from another user.

Minding Your Manners: Mastodon Etiquette

One mark of a Mastodon newbie is a post that starts "I hope this is OK. I don't want to get banned right away." This is especially true when the post isn't controversial. Although an instance administrator or moderator can kick you off a server for violating the instance's code of conduct, this rarely happens.

The tips in this section will help you to stay on the good side of the admin and be a good citizen of Mastodon.

Of course, if you're a troll or another type of disruptor trying to get yourself banned to discredit Mastodon or one of its instances, you're likely to get your wish. Hardly anyone will be sad to see you go, either.

Review the code of conduct

When checking out an instance in search of your home, don't skip over the code of conduct! Not all instances pattern their codes on mastodon.social's code of conduct.

To find an instance's code of conduct, look at the About page for the instance. Some instances also list *server rules*, which consist of a short list of the conduct to avoid.

TIP

Be aware that white supremacists, spammers, and other shady characters gravitate toward instances that define themselves as free speech zones.

If you have a question about something in the code of conduct, ask the administrator or moderator.

Be conversational

On Twitter, you're encouraged to gather followers and build a platform. Your follower count is displayed prominently on your profile page. Celebrities put out press releases when they hit a million followers. For the not-so-famous, you're still encouraged to stand out in the crowd. Some people make bold declarations and challenges. Others use sarcasm. Any helpful information often takes the form of a sales pitch.

This approach simply doesn't work on Mastodon. As we note in Chapter 4, what's important is joining in the conversation. When sharing something you found on the internet or in your feed, tell the community what you found interesting — or funny or disturbing — or explain what you learned. You have 500 characters to do so. The conversation shown in Figure 5-1 offers a great exchange of information.

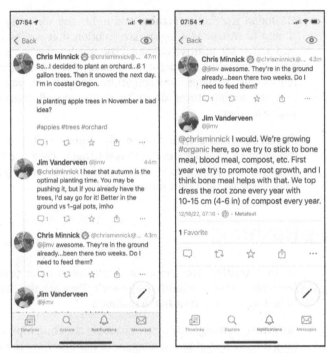

FIGURE 5-1: Good conversations are civil and often solve problems.

Respond to people who participate in a thread you started. You don't necessarily have to offer another comment. Sometimes just marking their post as a favourite will do the trick.

On Mastodon, almost all posts are from a human being. Respond to them as if you were talking on the phone.

REMEMBER

Build community

Mastodon's main purpose is community building. The avatars populating your Local tab are your virtual neighbors. Succeeding in Mastodon may require more of a mindset shift than a behavior change, but you should always remember why you came to Mastodon in the first place.

Successful virtual communities involve people who come together with a common purpose, whether it's learning, service, taking action to solve a problem, or just to have fun. Mastodon contains all these types of communities and more.

Mastodon isn't all sweetness and light, and we're not saying it should be. Anger is a legitimate emotion that can be channeled into action for positive change. (Too much anger isn't useful for anyone, including the perpetually angry.) When friction between individuals develops into something more toxic, you can start bringing temperatures down by reminding all concerned why you're all there.

Mastodon offers a virtual space where people all over the world can connect over whatever interests them. They can engage in conversation about their common interests and, potentially, change the world. Don't mess it up.

Using Hashtags

In an algorithm-free network, hashtags are the primary way posts get surfaced around the network. They are also the best way to be seen outside your home instance.

On Twitter, an algorithm can assemble trending topics by scanning the full text of posts, whether or not they have the # symbol. This is not true on Mastodon. So it's up to you whether people find the information or the conversation in your post. Help them out by including relevant hashtags.

When you make a public post to Mastodon, hashtags help others find your post. (Think of a hashtag as an index term in a nonfiction book.) You don't want to overload your post with hashtags, though. A 200-character post with 15 hashtags of dubious relevance will likely be blocked as spam.

TIP

No special procedure is needed to add a hashtag to a Mastodon post. Mastodon will turn any text starting with the hash symbol (#) into a clickable link, as shown in Figure 5-2, left. Sometimes, it will even suggest the hashtag while you're typing (try

#introduction). The link takes you to a page containing posts with that hashtag, as shown in Figure 5-2, right. Click the add icon to follow posts with this hashtag.

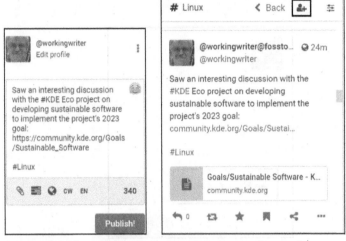

FIGURE 5-2: Adding a hashtag to your post lets more readers see it.

Everyone is an expert at something, whether it's their job, their education, their sports team, or their hobbies. Show off your knowledge and experience by following relevant hashtags. Answering a person's question on Mastodon not only helps that person but also helps you gain followers. Don't be arrogant and don't simply refer the person to, say, a list of frequently asked questions on your website.

For information on finding and following hashtags, see Chapter 3.

Enhancing Your Posts

You've covered the basics of posting on Mastodon. Now it's time to learn about the five ways to enhance your posts, which are found at the bottom of the Edit Post window, as shown in Figure 5-3.

Working from left to right, they are

>> Multimedia (paperclip)
>> Poll (three horizontal bars)

» Privacy guard (globe)

» Content warning (CW)

» Changing language (two-letter code of your primary language)

FIGURE 5-3: Use these icons to add different types of content to your posts.

The mobile app lineup of icons is shown in Figure 5-4:

» Multimedia (image)

» Polls (three vertical bars)

» Emoji (smiley face)

» Content warning (!)

» Privacy guard (globe)

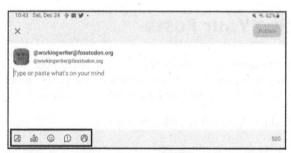

FIGURE 5-4: In the Mastodon app, click the emoji in the Edit Post bottom menu.

Let's walk through these options.

Illustrating your posts with videos and images

People gravitate toward social media posts that contain pictures or video. There's a reason why so many jokes about social media and the alleged obsessiveness of its users reference cat videos. Visuals enhance the story you're trying to tell, drawing people to read your post.

Photos and videos do not apply to your post character count.

You can also add emojis to a post. To do so in a browser, simply click the insert emoji icon in the upper-right corner of the Edit Post window to display the screen shown in Figure 5-5. In a mobile app, use the system's emoji keyboard.

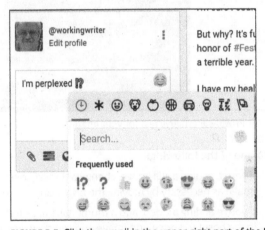

FIGURE 5-5: Click the emoji in the upper-right part of the Edit Post window.

To see a description of an emoji, mouse over it.

Adding a media file — photo or short video clip (up to 40MB) — from your computer to a post involves just a few steps:

1. **Click the paper clip icon at the bottom of the post.**

   When posting with the official Mastodon app in iOS or Android, use the standard method for opening an image.

2. **In the dialog that appears, select a photo or video clip from your computer.**

    On Apple iPhones, you can take a photo in the Mastodon app. Be careful, though, because the photo may exceed the 10MB posting size limit.

3. **Click the Open button or the Upload button to upload the image.**

4. **Click Edit in the upper-right corner of the image container.**

    The Edit Media window appears, as shown in Figure 5-6.

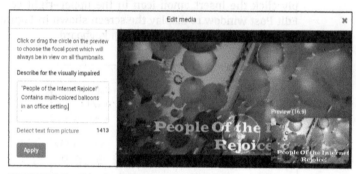

FIGURE 5-6: Provide alternative text for your image.

5. **Do one of the following:**

    • If you're adding a still image, change the focal point of the image as necessary by moving the bright circle.

    • If you're adding a video, add a thumbnail image to display in the post by clicking Add Image and selecting an image.

6. **In the box on the left, enter a full description of the image.**

    This text is called alt-text, for alternative text. Screen readers can process this text, giving visually impaired users a sense of what the image describes. You have 1,500 characters for describing the image, though most descriptions are much shorter. If the image also contains text, you can include some or all of that text in the description. If you click the Detect Text from Picture link, Mastodon will attempt to read and display the text from the picture in the alt-text.

    This step shows that you care about usability and accessibility, which are pluses for your credibility on Mastodon.

7. **To upload the image or video clip to your post, click Apply.**

   When the image or video is uploaded to the Edit Post window, the Mark Media as Sensitive box appears.

8. **To apply a content warning for an image, select the Mark Media as Sensitive box.**

Creating a poll

Curious minds ask questions. Polls are a way to get answers from your cohort of highly intelligent respondents: your followers!

Mastodon polls are anything but scientific, but they can be useful in learning something about your followers. They can also be just for entertainment, as shown in Figure 5-7. How you use a poll is up to you.

FIGURE 5-7: Mastodon polls don't have to be serious.

To create a poll on any subject, with multiple-choice answers:

1. **In the Edit Post window, click the icon with three horizontal bars.**

2. **Type your poll question.**

   If you want to maximize participation, ask respondents in the post text to boost the poll so others can respond.

3. **By default, respondents get two choices, but you can click the Add a Choice button to expand the range to a maximum of four answers, as shown in Figure 5-8.**

   In the mobile app, you can change the order of the choices by dragging the handles on the right side.

FIGURE 5-8: Mastodon polls allow you to select up to four response choices.

4. **Set the amount of time people get to respond to the poll.**

   By default, the response time is one day. Click the arrows next to 1 Day to choose from a minimum of five minutes to a maximum of seven days. In the mobile app, tap Duration to see these options.

5. **Click the Publish! button.**

Your poll stays active for the period you designate. People who respond will get updates on poll results in their notifications, with a final result when the poll expires.

Setting post privacy

Mastodon offers you the ability to restrict access to posts that you don't want the whole world to see. You can define these privacy settings while you're creating the post.

To the right of the paperclip icon, you'll see a tiny globe, indicating that this post will be publicly available to anyone in the fediverse. Click the icon to change who can see your post, as shown in Figure 5-9.

FIGURE 5-9: Privacy options let you choose who sees your post.

Your options are as follows:

» **Public:** This setting is the default. Anyone in the fediverse can see it, find it in a search query, and boost it to others to extend its reach.

» **Unlisted:** Anyone can see the post, but no one can find it in a search query. Followers (who can see everything you post, except those marked Mentioned People Only) can boost your post. This option is available in only in the web interface. This option is not available in the mobile apps.

» **Followers Only:** People who have already declared that they want to see everything you post are the only people who will see the post. Followers can't boost the post.

» **Mentioned People Only:** The post will be private or semi-private. "Meet me at Charlie's Bar at 7 pm. Cool things will happen." Recipients can't boost the post. The mobile apps call this option Only People I Mention.

You can combine post privacy settings and content warnings to define who can view and interact with every post you make.

Adding a content warning

You can add a content warning to any of your posts.

Your followers can identify their default preferences for handling posts with content warnings. See the section on setting your preferences in Chapter 3.

To add a content warning on your post text:

1. **In the Edit Post window, click CW (or ! in the mobile apps).**

2. **In the first field, add your warning text, as shown in Figure 5-10.**

FIGURE 5-10: Enter your public warning content in the top box and the sensitive content on the bottom.

3. **If you want to limit the audience, edit your Post Privacy settings.**

    See the preceding section for information on editing the settings. You can do this before or after adding the text.

4. **Click Publish!**

Your post appears with the text hidden behind a Show More box, as you can see in Figure 5-11. Your reader can click this box to see the content, and then click Show Less to hide the text again.

FIGURE 5-11: The recipient can view hidden content by clicking Show More.

Changing your language

Mastodon serves a global audience, and one way it acknowledges that is by providing the ability to publish posts in dozens of languages.

Changing the language in a post's settings allows the post to appear when others are searching for posts in the language you select.

TIP

You can't change your language in the mobile app. However, for the same effect, you can change your posting language by choosing Profile ⇨ Preferences ⇨ Other ⇨ Posting Defaults. Return to this page to restore your default language.

If you're multilingual and want to communicate with people in a language other than your default language, click the button with the two-letter abbreviation of your default language (for example, EN for English). A drop-down menu appears, so you can make your selection. You can also use the search box to type the name of the language in that language. In Figure 5-12, the language selected is Deutsch (DE), or German. Click the Translate link (Figure 5-13, left) to see what it means in English, as shown in Figure 5-13, right.

FIGURE 5-12: Click Translate to see what this German-language post says.

FIGURE 5-13: This post is excellent!

Your language choice applies only to the current post. Your Edit Post window returns to your default language for the next post.

WARNING

Changing the language in the Edit Post window does not translate your English post to the selected language.

Chapter **6**

Mastodon for Business

The path for businesses to becoming involved in Mastodon is being charted, and it looks a lot different from how businesses participate in traditional, commercial social networks. Without advertising and with overt marketing generally discouraged, setting up a Mastodon account for a business is potentially fraught with pitfalls. However, done right, having a business presence on Mastodon can become a valuable way to connect directly with your customers and build loyalty.

In this chapter, you discover some tips for dipping your business's toes into the fediverse, and you find out what other businesses are doing.

Doing Business in the Fediverse

With the explosion of popularity of Mastodon, and with many advertisers becoming wary of Twitter, marketers are chomping at the bit to get going on Mastodon. However, Mastodon is unlike other social networks. Think of participating as a business on Mastodon as a bit like visiting someone's home. Are you a traveling vacuum cleaner salesperson who comes by at dinnertime, or are you the plumber who unclogs the sink before guests arrive?

Taking it slow

The first thing to know is that you shouldn't be in a rush to get your business on Mastodon. With no obvious path to making money and with a userbase that's generally not welcoming of anything that smacks of commercial activity, carefully consider whether your business needs to be on Mastodon at all.

Deciding whether Mastodon is appropriate for your business

The users of Mastodon are a bit more technical (at least for now) than the users of other social media platforms. This means they might also be more savvy with where they look online for goods and services. No one, in our experience, is currently going to Mastodon to research things they want to buy.

However, if your business appeals to a more technical audience and is eager to get involved with helping loyal customers who are joining Mastodon, you may have a great opportunity to connect with people who have been turned off by Twitter and will see your presence on Mastodon as a rejection of what Twitter has come to represent to them (for example, extremist views, hostility, and prioritization of a billionaire's whims over sensible moderation).

Figuring out if your presence will help

There are other social media ecosystems besides Mastodon and Twitter, and some of them (including Facebook, Instagram, Pinterest, and TikTok) have more users than Mastodon and even Twitter. If your goal is to reach as many people as possible, Mastodon is not for you. If your goal is to meet your customers where they are, your business might be a good fit for the Mastodon community.

Discovering the kinds of businesses that thrive on Mastodon

If you have a loyal base of fans on Mastodon, you will get followers and you can interact with them. Examples of businesses that would be a great fit for Mastodon include

>> Sports teams
>> Non-profit or activist organizations
>> Newspapers and news websites

- » Book publishers
- » Podcasts
- » Television shows
- » Game developers
- » Companies known for supporting the open-source community

Monitoring what other businesses are doing

What are your competitors or similar businesses doing on Mastodon? Checking out whether they've joined, what they're doing on Mastodon, and how it's going for them may be a great way to gauge whether the time is right for your business to join too.

If you're searching for a particular business on Mastodon, be aware that anyone can create a user account on an instance using any name that's not taken on that instance. Generally, well-known companies will choose to either join a large instance (such as mastodon.social) or start their own instance (with their own domain name).

When searching for a well-known brand or company name, it's common to see several accounts that use that name. For example, Figure 6-1 shows just a few of the results that appear when you search for usernames with *Twitter* in them on Mastodon. To identify the official one, you'll need to look on the user's profile for a verified link. (Verified links are covered in Chapter 2.)

FIGURE 6-1: None of these is actually Twitter.

Knowing the Policies of a Server

Because Mastodon is not just one server run by one central set of rules, the rules about business accounts vary. Some servers strictly disallow business accounts. Some servers wouldn't be a good fit for business accounts even if they allowed them. Make sure to check the rules of any server before joining it.

TIP

If your business has the resources to do it right, your best bet might be to start your own Mastodon instance. We discuss that topic at the end of this chapter and in Chapter 7.

Find out whether business users are allowed

With thousands of separate instances making up the global Mastodon network, the differences in policies between them can be substantial. We know of at least one instance (Chris's home on Mastodon, Hachyderm.io) that has a well-thought-out corporate account policy.

Reading the rules

The Hachyderm Corporate Covenant (shown in Figure 6-2) is an excellent blueprint for anyone wanting to have a corporate account on any Mastodon server.

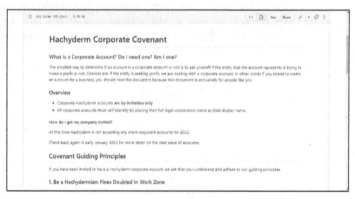

FIGURE 6-2: A sample corporate covenant.

It states that corporate accounts are by invitation only, and then proceeds to define what constitutes a corporation and the rules that corporations agree to when they create an account. These rules include the following:

>> Abide strictly by the rules of the server.

>> Be the corporation you want to work with.

>> Don't spam.

>> If your content becomes "invasive," you will be asked to move to a different server.

>> If you cause the moderators or administrator of the server problems, you'll be asked to leave.

You can read the Hachyderm Corporate Covenant at `https://github.com/hachyderm/community/blob/main/accounts/corporate-accounts.md`).

Joining a business-friendly instance

At least one Mastodon instance, corp.social (shown in Figure 6-3), is specifically welcoming to businesses and doesn't allow individual accounts. It's possible, however, that moderators of other instances will choose to limit posts from business-friendly servers, which will make them visible only to users who follow them.

FIGURE 6-3: The corp.social business-friendly instance.

Find out whether marketing is allowed

Twitter and other social media sites are big enough that corporations can post the same thing over and over in the hopes that it will reach additional people each time it's posted. Mastodon does not work this way. Repeatedly posting the same thing or a slight variation is considered spamming on Mastodon. If you do that enough, you'll start to have your account blocked, which will achieve the opposite of what you're hoping for.

Posts should be well-thought-out and personal, as described in the next section. This is not a job to delegate to an SEO professional or (much worse!) a bot.

Adjusting to Mastodon

Mastodon aims to be a friendly and authentic place. Even well-intentioned users who post nothing but puns are sort of a nuisance on Mastodon. And a corporate account that posts nothing but advertising taglines is seen as much more of a nuisance than even the non-stop pun guy.

Make it personal

New content is important. Does your company produce a podcast that's interesting and not just an ad? Post a link to it on Mastodon! Do you have tips that fans of your game might enjoy? That's welcome as well. Does the CEO want to explain why the company is no longer advertising with Twitter? That would be Mastodon gold!

Don't repeat yourself

If you post something once, you can rest assured that people who follow you will see it and will favourite or boost it if they want to. Repeatedly posting the same thing in different ways just annoys people who follow you, without reaching a new audience.

REMEMBER

Mastodon doesn't have an algorithm that rewards you for posting frequently. It's better to post infrequently or not at all than to post the same thing multiple times.

Ask for guidance

Server administrators and moderators love to be consulted about what's appropriate. Before you join a server, reach out to the administrator and ask about the server's policies regarding corporate accounts. Many smaller servers would be significantly harmed by a large corporation joining them because of the traffic (which translates to actual costs for the person running and paying for the server and bandwidth out of their own pocket). If your mere presence may contribute to an administrator having to shell out more money for hosting, or if having your account on that server might cause it to go down, you'll want to know before you join.

Making Positive Contributions

Being part of Mastodon can be rewarding for people as well as for companies. As a corporation, you may have access to resources that the administrators of a server do not have. Offering to help out can be a great way to become a welcome presence in the fediverse. A few ways you can help out include contributing for the following to your instance:

>> Money
>> Expertise
>> Equipment
>> Space

Contributing money

Joining Mastodon is always free. However, businesses especially should be conscience of the fact that most Mastodon instances are run entirely by volunteers. Most instance administrators have established a way to donate money to offset the cost of running the instance. Figure 6-4 shows one such donation page.

You can also donate to the Mastodon project itself, to help with the cost of running the original Mastodon server (at mastodon.social) and developing the Mastodon software, as shown in Figure 6-5.

For non-business users, donating is optional. If you're a business on a Mastodon instance, donating is optional as well, but making some sort of financial contribution is a good practice and will help make your presence on the instance more welcome.

FIGURE 6-4: Making a donation to your instance administrator.

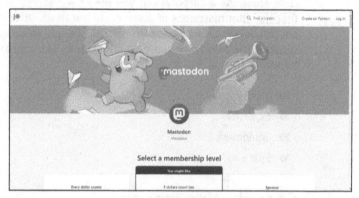

FIGURE 6-5: Making a donation to Mastodon.

On many sites, contributions of money can gain a business perks or branding. On Mastodon, it's best to consider donating to an instance because it's the right thing to do, rather than because you'll get your logo on a banner. If the instance has a policy of listing people and businesses who contribute, that's great, but it's not something you should ask for in return for your contribution.

Business contributions to an instance aren't necessarily listed, but those to Mastodon itself are listed, at https://joinmastodon. org/sponsors (see Figure 6-6). Businesses that want to become sponsors of the Mastodon project can go to https://sponsor. joinmastodon.org (see Figure 6-7) to learn more.

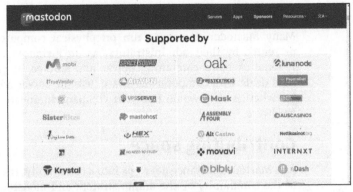

FIGURE 6-6: Viewing business sponsors of Mastodon.

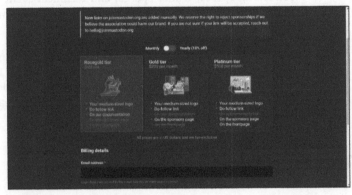

FIGURE 6-7: Becoming a Mastodon sponsor.

Contributing expertise

If your business deals with networking or computer programming, you may employ people or have resources that could be valuable to the administrator of the Mastodon instance you join. Keeping a busy server running and making sure that the database and server are properly configured is often not an easy job — especially for what might be just one person who decided to start an instance.

Commercial social media platforms have large teams of experts who monitor, configure, and maintain the computers that keep things running. If you can donate some of your time or the time of others at your company to help keep an instance running, it will be appreciated.

Contributing equipment

Many Mastodon instances run on physical computers owned or rented by their administrators. As an instance gets busier, it requires more computing power and storage. If your business deals with computer hardware, ask the server administrator whether they would like your company to donate computer hardware.

Contributing space

As a Mastodon instance develops into a community, the members of the community or the administrators and moderators might choose to get together offline. Businesses such as coffeeshops, coworking spaces, maker spaces, and technical event producers might consider offering space to relevant Mastodon instances so they can meet or be part of an offline event.

For example, a conference about the Linux operating system might consider reaching out to the administrator of the fosstodon. org Mastodon instance, which is an instance for people involved in free and open-source software, to donate meeting space or host an event for its users.

Running Your Own Server

Organizations that have the resources and expertise should consider running their own Mastodon instance, rather than creating an account on someone else's server. With your own Mastodon server, your Mastodon address can use your own domain name (for example, Microsoft might use microsoft.social) and there will be no doubt about which Mastodon account that uses the name *Microsoft* is the official one.

When you create your own server, you also get to set the rules about what is permissible on the server. If you're wary about continuing to do business on Twitter because its policies are unreliable and changing, one of the best things about Mastodon is that you have the opportunity to be in control of your own instance.

A Mastodon server can be set up in a day or two, but properly maintaining and administering a server can require a significant time investment. For more on setting up and running your own Mastodon instance, see Chapter 7.

Chapter **7**

Running Your Own Mastodon Instance

Perhaps you've mastered the basics of Mastodon and feel comfortable and knowledgeable. Or maybe you have some skills that might help the project.

In this chapter, you explore some options for getting involved in the Mastodon Project, whether it's owning or managing an instance, moderating conversations, or fixing bugs.

Building an Instance

Most people just engage with social media. They post, connect with friends, talk about stuff, and share funny memes.

But you're different. You want some control over the experience, deciding who's in, who's out, what the rules should be. In short, you want to be a social media mogul on a small scale. Mastodon and other fediverse-based services are the only platform where you can start and run an instance, with your own self-defined community. In this section, we show you how.

REMEMBER

When we talk about an *instance* in this chapter, we mean the computer running the Mastodon software. When we talk about a *server*, we mean the generic piece of computer hardware.

Hosting your own instance versus using a hosted instance

First, you need to make a decision. Will you build everything from the ground up with a self-hosted instance? Or do you want professionals to handle some of the technical steps to getting a Mastodon instance up and running while you focus on creating a perfect community space?

Following are the benefits of hosting your own instance:

» You're in complete control. As the owner of the instance, you decide what server to use, which internet service provider or cloud service will connect your server to the web, who to invite to participate in this adventure, and how much you want the instance to grow.

» If you're looking to establish a community around your business, the server can be housed in your building, with the company's name attached to the domain.

» You can define what kind of community you want, with its own rules and code of conduct for users.

You can host your own instance using a small-scale cloud service such as Linode (www.linode.com/marketplace/apps/linode/mastodon-server) or Digital Ocean (https://marketplace.digitalocean.com/apps/mastodon) or a giant such as Amazon Web Services or Google Cloud.

And here are the benefits of using a managed host for your instance:

» You don't have to handle all the technical details.

» For a single monthly payment, a hosted version of Mastodon handles internet connection and security, database backups, storage for user-uploaded media, and many other technical issues associated with building a community-based website.

>> When you have an idea for an improvement, technical advisors can tell you whether it's feasible.

>> You can focus on building your community.

In this chapter, we use the managed host Masto.host. Other managed hosts include Spacebear (https://federation.spacebear.ee), Ossrox (https://ossrox.org), IKNOX (https://iknox.com/products/mastodon-hosting), Elestio (https://elest.io/opensource/mastodon), and Weingaertner IT (https://weingaertner-it.de).

Understanding Masto.host

Masto.host, which is managed by Hugo Gameiro, from Portugal, is the largest and most well-known Mastodon managed host. It's a great place to get started with creating and managing your own Mastodon instance.

Before you choose managed hosting from Masto, review the terms of service at https://masto.host/tos. You'll get a fair assessment of how difficult Hugo's world has been since he took on this task, as well as a sense of the headaches you may avoid by not building your own server.

We walk you through the Masto.host setup process here. The process of getting set up with another managed host should be similar.

Getting started with Masto.host

When signing up with Masto.host, the first question in the setup process is whether you want to host on Masto's servers or your own. If you have a website or own a domain name that can serve as the home for your Mastodon instance, you might want to host your own server.

If you're buying a domain name expressly to host a Mastodon instance, choose a name to reflect your intentions for the instance. If you're going to host an instance on your personal or company website, think about putting Mastodon on a subdomain (that is, Mastodon.*mysite*.com). You might need to check with your website hosting company to confirm whether they support subdomains in your current setup.

If you want to host on your own domain versus hosting on masto.host, you should know enough about the domain name system (DNS) to point Masto.host to your domain. If you prefer not to dig that deep into internet standards, you'll likely want to host on Masto's servers.

Setting up your account and the server

In this section, you set up your MyMasto hosting account:

1. **Go to https://masto.host/tos.**

   You see the choices shown in Figure 7-1.

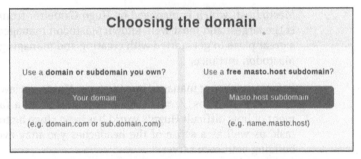

Choosing the domain

Use a **domain or subdomain you own?**

Your domain

(e.g. domain.com or sub.domain.com)

Use a **free masto.host subdomain?**

Masto.host subdomain

(e.g. name.masto.host)

FIGURE 7-1: Choose whether you want to use your own domain or Masto's.

2. **Do one of the following:**

   - *To host on Masto's servers:* Click the Masto.host Subdomain button. When asked, choose a name for the server domain.

   - *To host on your own domain:* Click the Your Domain button. Then enter the domain name (or subdomain) you want to use for the Mastodon instance. You'll receive custom instructions for redirecting the (sub)domain from your existing web host to Masto. You can do that after completing the signup steps listed here.

WARNING

Do not install Mastodon on www.*mysite*.com/mastodon if there is already content on *mysite*.com because Masto.host would replace your existing content with Mastodon. If you want Mastodon to co-exist with your existing content, choose a subdomain.

The screen shown in Figure 7-2 appears. This is the start of the setup of your Masto hosting account.

FIGURE 7-2: Set up a payment plan for Masto.host.

3. **Click Subscribe.**

4. **Enter your email address and credit card or PayPal information on this page and then click Enter.**

   The Set Password page appears.

5. **Enter a password (twice) for the My Masto account, as shown in Figure 7-3. Then press Enter.**

   This password lets you log in to and manage your hosting account. After you set up your password, Masto begins to set up your server and sends a Welcome email to the address you entered.

REMEMBER

If you're setting up on your (sub)domain, your server won't be accessible to users until you've redirected the site to Masto, as mentioned in Step 2.

Setting up an owner account on Mastodon

After you receive the Welcome email mentioned in the preceding section, a second email with the subject line "Mastodon server *domain*.masto.host is ready" follows when your server is accessible. This email asks you to set up a user account on the instance. To do so, follow these steps:

1. **In the "Mastodon server is ready" email, click the link to your new Mastodon instance.**

2. **Create a new user account on the instance.**

   This takes place in the Mastodon user interface. For details, see Chapter 2.

FIGURE 7-3: Setting your MyMasto hosting account password.

3. **When the new user account is registered, go to my.masto.host and log in with the password you set in the preceding section.**

    The account dashboard page appears, as shown in Figure 7-4.

4. **In the Hosting section, click your server name.**

    The Masto hosting page appears, as shown in Figure 7-5.

5. **In the Mastodon Server section, click the Change User Role button.**

    The screen shown in Figure 7-6 appears.

6. **In the first field, enter your Masto.com account password.**

    This is the password you set in Figure 7-3 (not your Mastodon user password).

7. **In the second field, enter the username of the Mastodon account you just created.**

8. **Click the New Role drop-down menu and choose Owner.**

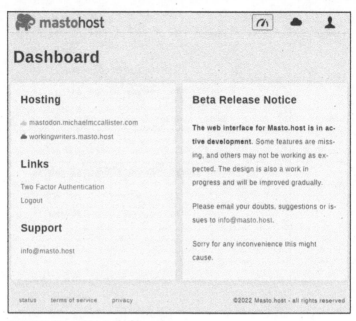

FIGURE 7-4: Your Masto dashboard page.

FIGURE 7-5: Your Masto hosting page.

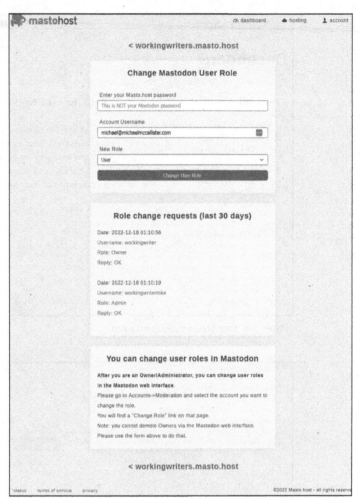

< workingwriters.masto.host

FIGURE 7-6: Define the owner of the Mastodon instance.

The account now has all the privileges of the server owner, up to and including terminating the instance.

TIP

Use an administrator account rather than the owner account for making administrative changes to your instance. Using the administrator account is safer because it has fewer privileges and can't make owner-level changes. Use the owner account for making only major changes.

Opening Your Instance to the World

In Chapter 1, we talk about creating a Mastodon instance for your friends and family. You're most of the way there. In this section, you discover how to bring friends and family to your instance and keep others out. You also find out how to invite the rest of the fediverse to your instance.

Adding (and limiting) users

When you log in to Mastodon on your instance as the owner or administrator, you'll see many more options on the Preferences page than an ordinary user sees. The owner has access to all the configuration settings for the instance. We focus on the essential tasks here.

REMEMBER

For the rest of this chapter, you'll be working in the Mastodon user interface, not in My.Masto (or another hosting environment). You must be logged in as an owner or administrator to see all the preferences discussed here.

After logging in, click the gear icon in the top-left column of the advanced web interface, between the Mastodon logo and the search box. If you don't see the gear, click the three-dots icon and select Preferences. You see the Appearance Preferences page, as shown in Figure 7-7.

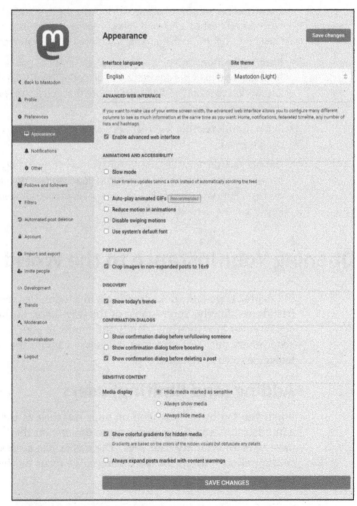

FIGURE 7-7: When you open Preferences, you're taken to the Appearance section.

Look for the Moderation and Administration tabs in the navigation bar on the left. This is where you'll be spending time while wearing your administrator hat.

Sending invitations

Administrators (admins) and owners can invite people to their instances. To invite people to your instance:

1. **Do one of the following:**
   - *If you're an owner:* In the navigation bar on the left side of Preferences, click Moderation, and then click Invites.
   - *If you're an administrator:* Use the Invite People option on the main menu.

2. The page shown in Figure 7-8 appears.

FIGURE 7-8: Click Generate Invite Link to send links to your Mastodon instance to friends and family.

3. **In the Max Number of Uses drop-down menu, select how many times this link can be used.**

   By default, there's no limit. If you're concerned about the link being spread around to people who may inundate your server, choose a smaller number.

4. **In the Expire After drop-down menu, select a link expiration date.**

   The default is Never. You can choose a range from 30 minutes to one week.

5. **If you want the invitee to automatically follow you when their account is set up, select the Invite to Follow Your Account check box.**

6. **Click the Generate Invite Link button.**

   The link is generated with the options you selected. You can now copy the link to include it in an email, on a website, or in a post on other social media platforms.

Requiring approval — or not

As we describe in Chapter 2, some instances allow new users to create an account automatically, while others require approval from an admin before joining the community. You should require approval if you want to limit the number of people joining your instance.

You decide the method you want to use in the Server Settings section of the Administration tab:

1. **In the navigation bar on the left side of Preferences, click Administration and then click Server Settings.**

2. **At the top of the page, click Registrations.**

   The screen shown in Figure 7-9 appears.

3. **In the Who Can Sign-Up drop-down menu, choose one of the following options:**

   - *Anyone Can Sign Up:* Displays the Create Account button on the signup page.

   - *Approval Required to Sign Up:* You'll be notified when someone signs up. When choosing this option, you can also select Require Reason to Join, which generates an edit box where applicants can explain why they want to join. You can then choose to approve or not.

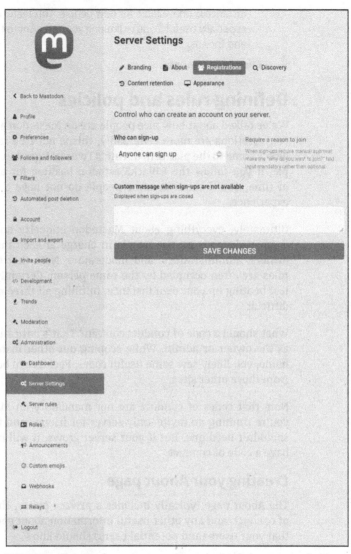

Server Settings

✏ Branding　📄 About　👥 Registrations　🔍 Discovery
↻ Content retention　🖵 Appearance

Control who can create an account on your server.

Who can sign-up

Anyone can sign up　⌄

Require a reason to join

When sign-ups require manual approval make the "Why do you want to join?" text input mandatory rather than optional

Custom message when sign-ups are not available
Displayed when sign-ups are closed

SAVE CHANGES

< Back to Mastodon
👤 Profile
⚙ Preferences
👥 Follows and followers
🝙 Filters
↻ Automated post deletion
🔒 Account
📤 Import and export
👥 Invite people
</> Development
📈 Trends
🔧 Moderation
⚙ Administration
🖥 Dashboard
⚙ Server Settings
🔧 Server rules
🖼 Roles
📢 Announcements
☺ Custom emojis
🔌 Webhooks
⇄ Relays
⬅ Logout

FIGURE 7-9: Choosing how users can sign up is an admin or owner decision.

- *Nobody Can Sign Up:* Only those invited by the owner or administrator can create an account. When you choose this option, the Custom Message When Sign-Ups Are Not Available edit box becomes active, so you can create a message for the signup page that explains why your instance is unavailable for new people. This feature is especially useful if you're limiting your account to family and friends.

4. **Click Save Changes.**

Defining rules and policies

We've talked about how nice people are on Mastodon and how the conversations are more civil. Sadly, this is not true for everyone who's come to the platform during #TwitterMigration. For example, if you follow the #BlackMastodon hashtag for any amount of time, you'll see that some people do not have a welcoming experience.

Ultimately, everything about Mastodon etiquette and codes of conduct depends on the people in charge at each instance: the owners, administrators, and moderators. Note that these three roles are often occupied by the same person. Certainly, if you're just booting up your own instance, fulfilling all three roles can be difficult.

What should a code of conduct contain? That's basically up to you as the owner or admin. While scoping out other instances for a home, you likely saw some useful codes. Feel free to borrow rules from those other sites.

Note that codes of conduct are not mandatory on Mastodon. If you're running an invite-only server for friends and family, you shouldn't need one. But if your server grows, it will be useful to have a code of conduct.

Creating your About page

The About page typically includes a privacy policy, the full code of conduct, and any other useful information about your instance that your users (and potential users) should know.

Following is the recommended information to include in the Extended Description section of the About page:

>> How you run the server

>> How you moderate posts on the server (this is where the code of conduct comes in)

>> How you fund the server

To build an About page, do the following:

1. In the navigation bar on the left side of Preferences, click Administration and then click Server Settings.

2. At the top of the page, click About.

3. In the edit box shown in Figure 7-10, type or paste the text of your extended description.

   You can prepare this text ahead of time, ideally in a text editor or word processor.

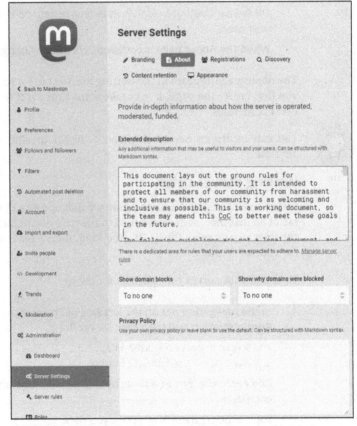

FIGURE 7-10: Extended descriptions can include the code of conduct.

4. **If you want to display a list of blocked domains:**

   a. *Under Show Domain Blocks, select who should see a list of blocked domains.*

   b. *Under Show Why Domains Were Blocked, select who should see why the domain was blocked.*

5. **By default, both items are set to the To No One option. Use the drop-down menu to select either To Logged-in Local Users or To Everyone.**

6. **If you have your own privacy policy, enter it in the Privacy Policy box.**

   Otherwise, Mastodon displays its default privacy policy to your users.

7. **When the About page is complete, click Save Changes.**

The About page appears in the Getting Started tab for users. When you first create the page, it may take some time for it to appear in users' timelines.

TIP

You can return to Server Settings ⇨ About to edit your page anytime.

Writing server rules

Server rules consist of a brief summary that appears in the About tab of an instance, sometimes next to the code of conduct. You can consider server rules the TLDR (too long, didn't read) version of the code of conduct.

To create your server rules, do the following:

1. **In the navigation bar on the left side of Preferences, click Administration and then click Server Rules.**

   The screen shown in Figure 7-11 appears. Each rule is displayed below the Add Rule button.

2. **Add each rule, one at a time, by clicking Add Rule after each item.**

   You can delete them at any point you want.

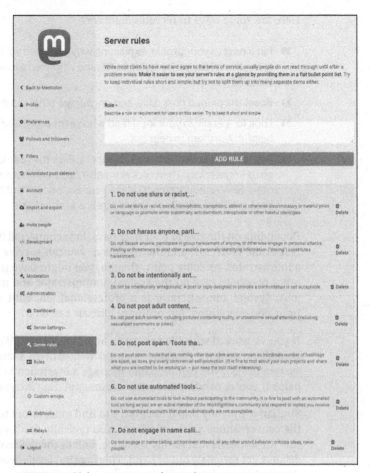

FIGURE 7-11: Make one server rule at a time.

Keeping your instance staffed

As an instance owner, part of your job is seeing that the system runs smoothly. At the beginning of your instance's life, that job likely belongs to you. However, someday you may not be able or willing to continue in that role, and you'll have to find volunteers to manage the system and its users.

Here are some ways to recruit volunteers:

» Pin a post to your profile, explaining why you need volunteers, the role(s) you're trying to fill, and who you're looking for to fill the role.

» Boost the pinned post daily, and encourage others to do so.

» Look for people in your local feed who seem active and helpful. Ask them to help out.

» Post an announcement to all members. Click the gear icon to go to Preferences. Then click Administration ⇨ Announcements ⇨ New Announcement. Make your announcement brief because it will appear on users' screens as they log in.

Depending on the size of your user base, how connected it is to the rest of the fediverse, and the workload, you can survive with one administrator. As some point, though, you might have to plan for other people to take on the system administrator and moderator jobs. System administrator is a professional job, with certifications and everything, but ordinary mortals can do it.

If you've worked with file servers in an office, or with WordPress, Drupal, or another database-focused web content management system (CMS), you can probably manage a Mastodon instance. It helps if you're comfortable with a command-line interface.

Similarly, you may find it necessary to find moderators to manage the conversations on the platform. You typically don't need to do this for a friends and family instance, unless the discussions and debates get seriously out of hand. Then a calm, level-headed family member can intervene.

You learn more about the duties of admins and moderators later in this chapter in the "Keeping Your Instance Running: System Administration" and "Enforcing the Rules: Moderation" sections.

Funding your instance

Have a look at the About page for your favorite instance. You'll likely see an accounting of the current financial situation of that instance. Here are some of the expense categories:

>> Hosting (by far the largest chunk)

>> Content Delivery Network (CDN) hosting, which speeds the delivery and display of media files on websites, including Mastodon instances

>> Server costs

>> Security software

Typically, once the expenses get larger than the owner can spare, they set up one or more funding mechanisms. Frequently you'll see requests for donations on Patreon, which was designed for recurring small donations, or PayPal.

Keeping Your Instance Running: System Administration

Mastodon system administrators should have some database management skills, especially with the PostgreSQL relational database. Maintaining the database also requires some skill at the command line, because you interact with Mastodon and PostgreSQL through their command-line interface, tootctl.

TIP For information on the Mastodon command-line interface, visit https://docs.joinmastodon.org//admin/tootctl.

Making backups

Data stored in four places on your instance's server has to be backed up periodically, to somewhere other than the Mastodon server:

>> **PostgreSQL database:** This database stores all the account, post, and follower information. Log in to the database and run either pg_dump or pg_dumpall to back up the data.

>> **Application secrets (including two-factor authentication details):** This information is in an environment variable file, .env.production, and needs to be backed up only once.

>> **User-uploaded files (mainly graphics in the form of avatars and attached media):** If these items are stored locally, you need to make copies of the public/system directory on the server. If you store user-uploaded graphics

on a Content Delivery Network (CDN) or a cloud service such as Amazon Web Services, you don't need to back up to a local directory.

>> **A Redis database:** The database file is stored in /var/lib/redis/dump.rdb and should be copied weekly.

Fighting spam with scripts

Mastodon checks the email address of every new signup against a spam domain blacklist. Spammers try to evade these controls by using multiple email servers, assuming that admins will be overwhelmed by the complexity of banning millions of domains.

Sometimes, however, all those domains resolve to a single IP address, and that's how you can fight back. For example, using the Linux dig utility with the domain name:

```
dig example.com
```

returns all DNS A records from example.com.

You might also try blocking certain IP addresses with the iptables and ipset utilities, although the process isn't foolproof. See https://docs.joinmastodon.org/admin/moderation/#blocking-by-ip.

Supporting user's discoveries

The administrator is the first point of contact for the user of any instance. Many users depend on the administrator to make their time on Mastodon as pleasant as possible.

In addition to answering questions, admins can make the first impression of Mastodon a good one. Mastodon's Discovery settings offer several tools to enhance a user's early experience.

Go to Administration ⇨ Server Settings ⇨ Discovery. Then select the Enable Trends check box, which (as on the birdsite) identifies the posts, hashtags, and news stories gaining traction on your instance. Leave the Allow Trends without Prior Review check box deselected. Doing so gives you some control over what appears in this list. For example, if there's a rush of posts on topics that aren't relevant to the kind of community you want to build, you can hide them from the list.

Let every visitor view the public timeline without having to create an account. When new people find out how interesting your users are, they may just sign up. Select the Allow Unauthenticated Access to Public Timelines option to allow this.

A long time ago, Tom Anderson, the founder of MySpace, realized that if his network of online friends was to succeed, everyone who showed up and created an account should automatically get a friend. MySpace Tom was born, and MySpace was the place to hang out online, at least until Facebook opened its doors to the general public.

Mastodon instances can recommend accounts to new users. Whether that's an admin or an active, smart user who can help the newbies is up to you. On this page, add users to the Always Recommend These Accounts to New Users option.

The last option on the Discovery page, Enable Profile Directory, allows you to publish a list of all users who agreed to be discoverable.

Enforcing the Rules: Moderators

Moderators aren't necessary for small groups of people who know each other. But if you plan on growing beyond that small circle, a moderator is critical to keep out racist hordes and other undesirables who might invade.

Moderators are also responsible for fighting spam by blocking suspicious email domains and identifying appropriate IP rules.

Reading reports

Much of a moderator's time is spent following up on reports sent from users of server rule violations.

If a user spots a violation in a post, they can flag the post for the moderator's attention by clicking the three-dot icon at the bottom of the post and selecting Report *username*. The reporter then selects which server rules the post violates.

The report generates an email notifying moderators (you) and administrators of the report. When you receive the email, click the link to the report. The report page, which is shown in Figure 7-12, gives you all the details.

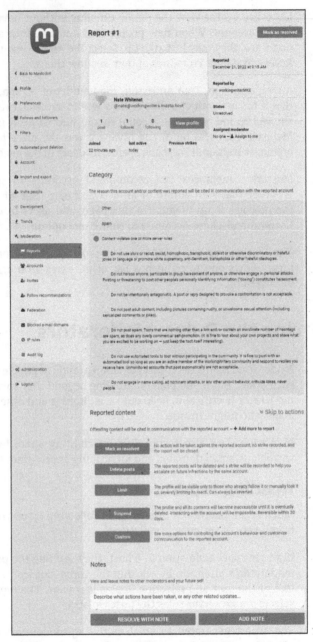

FIGURE 7-12: User reports enable you to take action to keep the instance safe.

As moderator, you may impose a variety of disciplinary actions:

>> **Mark as Resolved:** Closes the report and you can't edit it further. Wait until you know you've taken all the steps to deal with this violation. Add a note to the resolution to explain your reasons for the action(s) taken.

>> **Delete Posts:** Removes the post from the local database and requests other federated instances to remove it as well.

>> **Limit:** Keeps the post available only to the user's followers, not in the local feed. Limited posts can show up in a search and when the user is mentioned in someone else's post. A post limited on your instance could appear on other federated servers because the limit isn't necessarily communicated to other instances.

>> **Suspend:** Removes the user's posts, uploads, followers, and other data from public view. This penalty is the most serious because the user essentially disappears from Mastodon. The user has 30 days to work to restore themselves in the moderators' good graces. If the issue is not resolved at the end of the 30-day period, the account is wiped out of existence. Admins also have the ability to remove the user's account data before the 30-day negotiating period ends.

>> **Custom:** Displays a set of lesser actions, along with Limit and Suspend. You can choose one of these options from this menu:

- **Send a warning:** Sends an email to the offender.

- **Freeze:** Prevents the user from logging in or posting to their account but preserves its existence. Moderators can restore the account at any time, at which point the user can resume their activity.

- **Sensitive:** Labels all media files posted by the account as sensitive. This option is helpful if a user persists in posting sensitive material (images or video) without a content warning.

All discipline is in the moderator's hands, but all reports are saved permanently in the Reports area.

Fighting spam

A lot of spam posters are prevented by admins and the Mastodon software itself (see "Fighting spam with scripts" earlier in the chapter). If someone slips through and spam is reported by a user, you can add the spammer's domain to your instance's blacklist:

1. **In the navigation bar on the left side of Preferences, click Moderation and then click Blocked E-Mail Domains.**

   The current list of blocked domains appears.

2. **Click Add New.**

3. **Enter the suspected domain name or the MX record included in email from that domain.**

4. **Click Resolve Domain.**

   Both the domain name and its associated MX record are blocked. The server reports on how many sign-up attempts were prevented in the last week.

Disciplining other sites

If it becomes clear that many of the violations of your instance's code of conduct come from another federated instance, moderators can block that instance. As with individuals, there are degrees of discipline your instance can implement against another site.

Moderators can

>> **Reject media files from the designated site.** Media files include avatars, headers, emojis, and media attachments.

>> **Limit past and future accounts from the designated site.** As with limited individual accounts, posts from a limited site's accounts can show up in a search or when the user is mentioned in someone else's post.

>> **Suspend past and future accounts from the designated site.** In this case, no content from the suspended site aside from account names will be stored at the local site.

Contributing to the Mastodon Project

You don't have to be able to code, much less be a software developer, to contribute something valuable to the Mastodon Project. In this section, we describe just a few things you can do to help.

Reporting bugs and making suggestions

Have you run into a problem, large or small, using Mastodon? Got an idea for a feature that's missing in Mastodon? Let the developers know about it!

Start by contacting your instance's administrator. If you're asking about a problem, there may be a setting to be fixed or an issue with the server. The admin may offer you advice on dealing with the problem as well.

You can also take your question, issue, or suggestion to the Mastodon Issues list on GitHub at https://github.com/mastodon/mastodon/issues. Use the search box to enter your request and perhaps find an existing report.

You may also want to search the Mastodon Discussion area at https://github.com/mastodon/mastodon/discussions. This area recently replaced Join Mastodon's Discourse channel.

You need to create a free account on GitHub to post a new bug or discussion topic. Creating an account opens a communication channel for developers to keep you posted on progress toward resolving the issue.

Preparing documentation

A lot of good advice on using Mastodon is shared on the platform itself. If you really want to know how Mastodon works, check out the official manual at https://docs.joinmastodon.org.

At the bottom of every topic page, you'll see the date when the topic was last updated. To the right is an Improve This Page link to the GitHub page for the Mastodon documentation project.

You can't edit the page directly, but with a GitHub account, you can make changes for the team to review. If your contribution is accepted, it will be added to the page.

The Mastodon documentation is composed in a format called Markdown. This is a plain-text format with simple tags that can be published on the web.

TIP

Many resources sheets are available to help you learn Markdown. Here's one from GitHub: https://docs.github.com/en/get-started/writing-on-github/getting-started-with-writing-and-formatting-on-github/basic-writing-and-formatting-syntax.

Getting involved with your instance

If you're part of a medium-to-large instance, you can volunteer to help out. These types of communities always need volunteers for the following:

>> **Admin:** Those with database or command-line experience

>> **Moderator:** Level-headed people with the ability to drop the hammer on violators of the code of conduct

>> **Blogger:** People who want to write posts for the instance's blog, communicating with users and the world at large

Chapter 8

Ten Tools that Integrate with Mastodon

Mastodon is based on the open-source ActivityPub protocol. Unless you're planning to write your own app that integrates with Mastodon, you don't need to know anything more than that ActivityPub is freely available to anyone to do with as they please.

Because it runs on a free and open standard, the official Mastodon desktop app and the mobile app are not (by a long shot) the only ways to use Mastodon. In fact, most long-time users of Mastodon will tell you that the official Mastodon app is currently inferior to many of the apps developed by third-party programmers. But you don't have to take anyone else's word for it, because nine of the apps listed in this chapter cost less than a beer at the local pub.

In this chapter, you discover ten alternative apps that you can use to log in to your home server and interact with your friends and followers in the fediverse. Each app has a slightly different way of doing things, so which you use comes down to personal preference.

TIP

We listed the URLs for each app so you can easily find out more about them and try them out. For the mobile apps, you can also just search for them in the Apple App Store or in Google Play.

Metatext

Metatext (`https://metabolist.org`), shown in Figure 8-1, is a free iPhone and iPad app for Mastodon.

FIGURE 8-1: The Metatext app.

One nice feature of Metatext is its large Toot icon in the lower-right corner, which makes creating a new post quick and easy. The top navigation lets you choose between the three timelines (Home, Local, and Federated), and the Explore screen shows an attractive view of currently trending topics. The iPad tablet interface for Metatext is just as easy to use as the iPhone version and adapts well to portrait mode and landscape mode.

tooot

The tooot app (https://tooot.app) is a free mobile app for both iOS and Android devices. The app, shown in Figure 8-2, looks very similar to Twitter.

FIGURE 8-2: The tooot app.

The tooot app is attractive and has an average rating of 4.3 in Google Play. It was originally designed for Chinese users, and may be the best option for users in China.

Whalebird

Whalebird (available at `https://whalebird.social`) is a downloadable Mastodon client that runs on macOS, Windows, and Linux desktop computers. Its UI, shown in Figure 8-3, is similar to that of the Slack app.

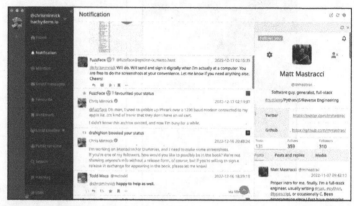

FIGURE 8-3: The Whalebird desktop app.

Some of the benefits of Whalebird versus a standard web browser user interface are that it can send desktop notifications, it has built-in shortcuts for common tasks, and it has many different color schemes you can choose.

Subway Tooter

Subway Tooter (`https://github.com/tateisu/SubwayTooter`), shown in Figure 8-4, is an app for Android phones and tablets for Mastodon. It features multiple columns that can be rearranged, support for multiple accounts, the ability to favorite, boost, and reply across different accounts, and support for Mastodon's latest features.

FIGURE 8-4: Subway Tooter for Android.

Tusky

Tusky (https://tusky.app), shown in Figure 8-5, is a Mastodon app for Android. It's currently one of the most popular Mastodon app choices (with over 100,000 downloads from Google Play). One of the best features of Tusky is that it's highly customizable.

FIGURE 8-5: Tusky for Android.

TheDesk

TheDesk (https://thedesk.top), shown in Figure 8-6, is a free desktop Mastodon client that provides functionality similar to other social media management apps such as Tweetdeck (for Twitter) and Hootsuite (for Twitter and various other services). You can manage multiple Mastodon accounts by using multiple highly customizable columns. The app also has some other unique features, including a text-to-speech feature and integration with several music services, including Spotify, Apple Music, and Last. fm. With a connected music service, you can include what you're currently listening to when you post.

FIGURE 8-6: TheDesk Mastodon desktop client.

TheDesk is available as an installable app for macOS, Windows, and Linux.

Toot!

Toot! (https://apps.apple.com/us/app/toot/id1229021451) is an iPhone and iPad app that currently costs $3.99 to download from the Apple App Store. If you want a beautiful app that's easy to use, up-to-date with the latest Mastodon features, and fast, the price may well be worth it.

Toot!, shown in Figure 8-7, has animation, the easiest posting experience of any of the Mastodon apps that we've tested, and support for multiple accounts.

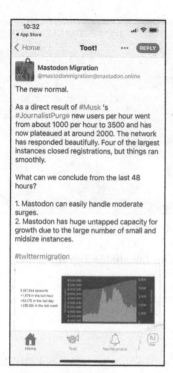

FIGURE 8-7: Toot! for iOS.

Tootle

Tootle (https://apps.apple.com/us/app/tootle-for-mastodon/id1236013466) is another free iOS Mastodon client app that works well on both iPhones and iPads. One notable feature of Tootle is the capability to store draft posts. It also features support for streaming services and for customizing both the theme of the app as well as your timeline.

Tootle is shown in Figure 8-8 running on an iPad.

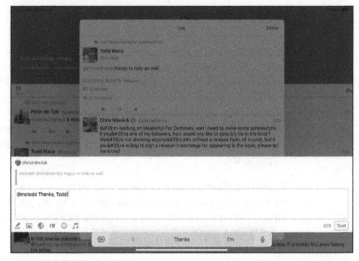

FIGURE 8-8: Tootle for iOS.

Halcyon

Halcyon (https://halcyon.cybre.space) is the most Twitter-like Mastodon client we've seen. The web app, shown in Figure 8-9, will look familiar to anyone who has used Twitter. Custom emojis don't seem to work currently, so you may not be able to get your blue check mark, at any price, while using Halcyon. But if you're looking for a familiar interface, this is the one for you.

FIGURE 8-9: The Halcyon client.

Hyperspace

Hyperspace (`https://hyperspace.marquiskurt.net`) bills itself as the "fluffiest client for the fediverse," whatever that means! Hyperspace, shown in Figure 8-10, can be downloaded and run as a desktop app on Windows, macOS, or Linux. In addition, a web version runs in a web browser. If you're technically inclined, you can download the web files and run your own Hyperspace server or even download the source code to build your own version of Hyperspace.

Hyperspace's interface is clean and attractive. You can customize it with ten themes, and you can choose the Show More Posts option in the preferences to view posts as a grid rather than in the default feed view.

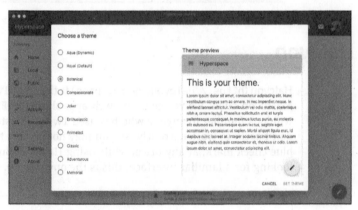

FIGURE 8-10: The Hyperspace app.

» BookWyrm, Inventaire, and Lemmy

» Owncast and Pubcast

Chapter 9
Exploring Other Fediverse Platforms

When you get comfortable in Mastodon, you might want to explore other parts of the fediverse. In this chapter, we provide a sampling of services available. All have some interaction with Mastodon, and the signup process is typically similar as well.

TIP

See even more fediverse platforms at Per Axbom's "The Many Branches of the Fediverse" at https://axbom.com/fediverse/.

PeerTube

PeerTube (https://joinpeertube.org/en) is a federated video-streaming platform for sharing files among its members and the public. Anyone can create a PeerTube instance to post videos. Each instance connects to the others and to the JoinPeerTube site.

If you just want to watch videos on PeerTube, visit its site (shown in Figure 9-1) and click the Browse Content tab at the top of the page. In the search field, type a topic, and you'll get results divided into videos, channels (devoted to your topic), and playlists (sets of videos curated by other users). Each result tells you who posted the video, what PeerTube server hosts the clip, and other details.

FIGURE 9-1: PeerTube offers video producers and fans an algorithm-free experience.

Although you may not see the latest official video from Taylor Swift, you'll have a chance to spend hours enjoying algorithm-free video. And PeerTube will not point you to some other clip that eventually leads you to extremist propaganda.

If you're a video producer, sharing your work on PeerTube, like Mastodon, requires you to choose an instance to host your content. The server page helps you find a community to join. Some PeerTube servers focus on hosting livestreams, and others offer a mix of recorded videos and livestreams. Several mobile apps show PeerTube videos as well.

Discourse

Discourse (www.discourse.org), shown in Figure 9-2, powers thousands of online communities, from Canadian Football League fans to Zoom technical support. Even Twitter developers use this fediverse resource to discuss issues and problems.

Here, every instance is a separate community. There isn't much direct interaction between instances, though one could see the potential for collaboration between Ubuntu Linux support and developer communities.

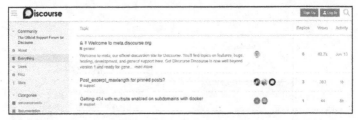

FIGURE 9-2: Get answers, technical support, and discussion in a Discourse community.

Pixelfed

Pixelfed (https://pixelfed.org) is a federated photo-sharing site that offers all the features of Mastodon (choose your home instance, sign the code of conduct, comment, and the like). However, the focus at PixelFed is on images, so it might remind you of Snapchat or Instagram.

You can join PixelFed directly from Mastodon as follows:

1. Go to https://mastodonservers.net/server/ 1077-pixelfed.

2. Click Join This Server.

   The Pixelfed screen appears.

3. Click the Register link below the Account Login form.

4. Fill out the Register a New Account form and then click Register.

5. Click Send Confirmation Email.

6. Open the confirmation email and click the Confirm Email button.

   Within seconds, your home page appears, as shown in Figure 9-3.

You can now start following people on the service to fill up your feed.

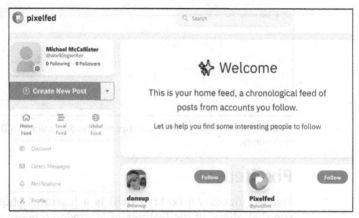

FIGURE 9-3: The home page at Pixelfed lets you start following people right away.

Friendica

If you've decided to avoid every corporate social network, Friendica (https://friendi.ca) might be your escape from Facebook. The list of similar features includes:

>> You can post status updates of unlimited length to share with your selected group, or everyone, or somewhere in between. Posts on Friendica appear also in Mastodon.

>> You can like a post, but you can also dislike a post (which you can't do on Facebook or Twitter).

>> Your Friendica posts can travel anywhere in the fediverse, and you can also email any post to anywhere outside the fediverse.

>> On Friendica, you can change the look and feel of your news feed.

The Quick Start feature shows you how to adjust your settings and start posting. A simple text editor with basic formatting buttons makes posting simple. Tabs let you define who can see your post, and a preview window (see Figure 9-4) shows you how the post will look on your profile page.

FIGURE 9-4: After writing your first post, check the Preview tab in the editor before sharing.

Funkwhale

Funkwhale (https://funkwhale.audio) is a music-sharing service that enables you to share the music you create with selected other people on the service, as shown in Figure 9-5.

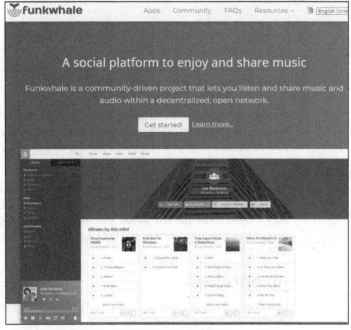

FIGURE 9-5: Find and share music and podcasts at Funkwhale.

You can also upload your digital music collection to a private library, allowing you to retrieve your music anywhere you have access to the web. Depending on your instance server, you can upload and store up to a gigabyte of audio content.

Funkwhale also supports the creation and distribution of podcasts.

BookWyrm

BookWyrm (https://bookwyrm.social) is a network for talking about books and writing and reading book reviews. You can search for books by keywords, or you can add books by scanning their barcode. If a book doesn't currently exist in BookWyrm, you can add it, as shown in Figure 9-6.

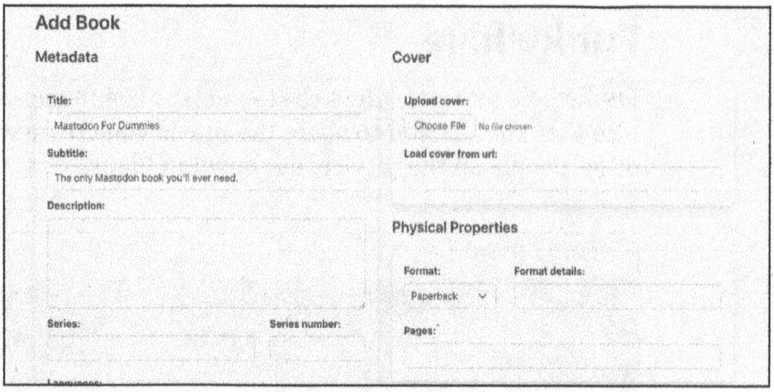

FIGURE 9-6: Adding your books to BookWyrm.

Similar in function to Amazon's Goodreads, BookWyrm has a feature that lets you track books you're currently reading, books you've read, books you stopped reading, and books you want to read.

You can also create lists of books. Members can share their lists with other members and set a list to open (anyone can add to the list), closed (only the list owner can add to the list), group (only people in a group can add to the list), or curated (anyone can suggest books).

Inventaire

Inventaire (https://inventaire.io/), shown in Figure 9-7, is another book-related network, but it's focus is on lending and borrowing books.

FIGURE 9-7: Borrow and lend books on Inventaire.

Once you're signed up, you can add a book to your inventory by searching for it, scanning its barcode, entering its ISBN, or by importing a file you can export from Goodreads or several other sites.

After you add some books, you can set your location and find other members near you and browse their library. If a user has a book you want to read, you can borrow it from them.

Lemmy

Lemmy (https://join-lemmy.org/) is a federated link aggregator. A link aggregator is a website where the main point of the site is for people to share links to websites. A federated link aggregator is a network of separate websites (like Mastodon) that can communicate with each other but can each have their own personality.

Similar to Reddit, Lemmy has a timeline where users can post links and other users can comment on them. Each instance has its own focus and members of every instance can access links from other instances as well.

Inside an instance, users can create communities for discussing topics. You can subscribe to any communities you like to add posts from those communities to your feed.

Figure 9-8 shows the page on Lemmy for finding and subscribing to communities.

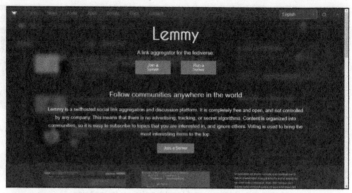

FIGURE 9-8: Share links and join communities with Lemmy.

Owncast

Owncast (`https://owncast.online/`) is a free, open-source, live video and web-chat server. Similar to Twitch, Owncast enables you to watch live video feeds from other users and chat as you watch. You can also integrate Owncast with popular broadcasting software to do your own live streaming. Unlike Twitch, however, you're also completely free to download your own copy of Owncast and operate your own server.

Pubcast

Pubcast (`https://github.com/pubcast/pubcast`) is an experimental distributed podcasting platform. As of right now, it's still in development. However, because it's open source, anyone is free to download it and try it out. If you have coding skills, you can even contribute to its development.

Index

About the Authors

Chris Minnick is a prolific writer of books about computers as well as of fiction. He's also an enthusiastic learner, amateur musician, and a winemaker. He lives in Astoria, Oregon, where he enjoys walking in the rain and mowing the grass when it's not raining.

Michael McCallister writes about the open web, open source, WordPress, and all things Linux from his home in the suburbs of Milwaukee, Wisconsin. He is also a technical writer senior for FIS. He loves baseball, but can't hit a lick.

Dedications

To everyone who works to figure out better ways to do things.

— Chris Minnick

To my family: Jeanette for being herself, Bev and Bob for their support, and the next generation, Hannah and Ben, who lead the way.

— Mike McCallister

Authors' Acknowledgments

Chris Minnick would like to thank the following: Jill and Chauncey for putting up with all the crazy early morning and late night typing; Michael for being my co-author and putting up with my hyperactivity and excitement to get this project done; Susan Pink for being just the perfect editor and a joy to work with; and Guy Hart-Davis for his insightful, thorough, and always correct technical editing.

Michael McCallister echoes Chris's thanks to Susan and Guy for their efforts to make this book the best it could be, occasionally facing a recalcitrant author. He thanks Chris, who was a joy to work with, for his enthusiasm and the occasional screenshot. He thanks Jeanette for her support and adapting to changing schedules. Finally, many thanks to the Mastodon communities that kept the lights on, waiting for the #TwitterMigration. A revolution begins when some people recognize the need for one. It succeeds when the need becomes obvious to everyone.

Publisher's Acknowledgments

Executive Editor: Steve Hayes

Senior Managing Editor: Kristie Pyles

Project Editor: Susan Pink

Copy Editor: Susan Pink

Technical Editor: Guy Hart-Davis

Proofreader: Debbye Butler

Sr. Editorial Assistant: Cherie Case

Production Editor: Saikarthick Kumarasamy

Cover Image: ©Nobumichi Tamura/ Getty Images